Easy-Up Antennas
for Radio Listeners and Hams

Easy-Up Antennas
for Radio Listeners and Hams

Edward M. Noll

HOWARD W. SAMS & COMPANY
A Division of Macmillan, Inc.
4300 West 62nd Street
Indianapolis, Indiana 46268 USA

FIRST EDITION
SECOND PRINTING—1989

International Standard Book Number: 0-672-22495-X
Library of Congress Catalog Card Number: 88-60042

Acquisitions Editor: *Greg Michael*
Illustrators: *T.R. Emrick and Craig Downey*
Cover Design: *Linda Simmons*
Composition: *Shepard Poorman Communications Corp.*

Printed in the United States of America

Trademark Acknowledgments

Contents

Preface

This handbook is a practical antenna book for the radio listener, radio amateur, and prospective radio amateur. All of the basics and some of the more advanced antennas are covered. Antennas are included for

- Shortwave Broadcast (SWB)
- FM Broadcast
- Medium-Wave Broadcast (MW)
- Long-Wave Broadcast (LW)
- Utility, Marine, and Special Services (UTE)
- Scanner Listeners
- Ham Radio

All of the designs are easy to construct, erect, and put into operation. Complete details are given. They are inexpensive to build and they stay up. Best of all—They Work!

This handbook is divided into two parts. Part 1 (Chapters 1 through 11) is for the radio listener and Part 2 (Chapters 12 through 16) is for hams. However, the contents of both parts are of value to listeners and hams alike and the entire book should be read through one time regardless of your special interest.

A special feature of this book is the detailed antenna dimension tables at the end of each part (Chapters 11 and 16). Dimensions for all of the basic antenna types are given for selected frequencies in each of the bands assigned for each radio service. Both the American (feet and inches) and the metric (meters and centimeters) dimensions are included. These tables can also be used to fabricate some of the more advanced antennas. Other tables in the book provide sets of constants that are useful for making dimension conversions and in calculating the antenna length for specific frequencies not in the tables.

Please read Topics 1-1 and 1-2 (Chapter 1) to obtain more information about the contents and features of *Easy-Up Antennas for Radio Listeners and Hams.* Topics 2-1 through 2-5 (Chapter 2) also contain much useful general information regardless of the type of antenna you plan to construct.

Each antenna presented in this book has been erected and tested by the author, except for the few commercial antennas mentioned in the text. Several of the latter were also purchased and tested. I thank Don Herrington for his painstaking and useful technical editing of the manuscript.

A number of the antennas covered in this book first appeared in print in my column "Better Signals" in *Popular Communications*. Thanks to the Editor, Tom Kneitel, for permission to reprint these articles. Also thanks to Larry Steckler, publisher of *Hand-On Electronics* for permissions related to articles of mine that appeared in this magazine. Refer to important addresses at the end of Topic 1-6 and in the Appendix for further information.

Edward M. Noll

Part 1

Radio Receiving Antennas

1. *Radio Services and Book Plan*
2. *Basic Antennas and Their Construction*
3. *Antenna Tests and Comparisons*
4. *Vertical Variations*
5. *Dipole and Longwire Variations*
6. *Directional Longwire and Beam Antennas*
7. *Special HF Antennas*
8. *Medium- and Long-Wave Antennas*
9. *Confined Space and Indoor Antennas*
10. *VHF/UHF Antennas*
11. *International Antenna Dimension Tables for Radio Listeners*

Chapter 1

Radio Services and Book Plan

WHEN you put up an antenna you throw out a lifeline between your receiver and the thousands upon thousands of radio signals that traverse your receiving site. Your antenna is the link between the great invention radio and the great natural gift electromagnetic wave propagation. The latter path was here eons before people walked the earth. I embrace both in awe and with respect.

1-1. Introduction

In reception you use your antenna to extract the feeble radio energy from the propagation medium which is able to convey the electromagnetic waves (radio waves) radiated from the transmitting antenna to your location. Part 1 of this book covers receiving antennas for the frequency range between 100 kHz (kilohertz) to almost 1000 MHz (megahertz). An antenna can also be used to link the radio energy generated by a transmitter to the propagation medium. Radio amateurs or "hams" have been doing this since the inception of radio. Part 2 of this book covers the ham radio antennas which provide an exchange of energy between two points. Actually the very same basic antenna types can be used for both transmission and reception. Hence, it is important that the radio ham or prospective ham read both Parts 1 and 2. Likewise the strictly radio listener should read both parts because ham-radio listening has become increasingly popular with the advent of the addition of sideband receivers capability to many general-purpose shortwave receivers.

The antennas covered in both parts of this book are effective and low cost. They can be built with few complications using readily available parts. In fact many of the parts can be purchased at bargain prices at hamfests, flea markets, and discount parts houses.

This is not a book of high-cost sophisticated applications. All the antennas for HF (high frequency), MF (medium frequency), and LF (low frequency) applications are strictly homemade do-it-yourself wire-type antennas. Commercial VHF/UHF antennas are also covered because it is often more economical to buy than to build antennas for this frequency range. However, several homemade types are presented too. Special consideration is given to antennas for the novice ham as well as the technician, and general class ham licensees. Of course more advanced styles are also covered in both Parts 1 and 2. They include beams, phased elements, and loaded antenna types.

1-2. Book Plan

Easy-Up Antennas for Radio Listeners and Hams has been arranged for convenient use. Part 1 on receiving antennas satisfies the primary interests of the radio listener only, while the transmit/receive antenna information of Part 2 is for the radio ham. However, to save space the essential fundamentals and basic construction procedures for both receive and transmit antennas are detailed in Chapter 2. This is an all-important chapter and everyone will do well to read it in preparing themselves for the most effective and efficient use of the book. Chapter 12 is an additional basic chapter for radio hams and prospective radio hams. In all cases it is helpful for everyone to read Chapters 1, 2, and 3 before becoming involved in the details associated with chapters of special interest. Chapter 1 gives you some additional introductory material and guidance such as band frequencies, time considerations, and sources of information. Chapter 3 includes important antenna test guidance. Reference antennas are so very important and deserve your consideration. Some additional ham radio back-up test data is to be found in Chapter 12.

Two sets of detailed antenna dimension tables are included. Dimensions for receive antenna for the radio listener are given in Chapter 11. Ham radio antenna dimensions are given in Chapter 16. The various constants and tables that are of assistance in designing antennas for almost any frequency are included in Chapter 11. The information here can also be used in association with the ham antenna information of Chapter 16. Many of the antennas described in the book show dimensions when cut for a particular band. The dimension tables can be used to find lengths for other bands. In addition antenna dimensions for other radio services can be calculated easily using the various constants given in Chapter 11.

Part 1 includes chapters on dipole variations, vertical variations for those with limited space, and (for the more daring) beams, longwires, and other special types and configurations. Indoor, mobile, and portable antennas are covered in appropriate chapters. Three special chapters of Part 2 offer antenna guidance to the new novice or prospective novice ham, new technician licensee, and new general-class licensee.

You have gotten this far in the book; why not carry on to the end of the chapter? Make a special effort to read Chapter 2 because it will help you better understand the antenna constructions and arrangements that follow. In later chapters you are sometimes referred to Chapter 2 to renew information.

1-3. Frequency Bands and Radio Services

Radio listeners and avid long-distance listeners (DXers) are to be found worldwide tuning their receivers freely over a frequency spectrum from less than 100 kHz to almost 1000 MHz. Unfortunately there are unenlightened countries that suppress radio listening. However in many of these countries clandestine listening is widespread or on the rise. Radio waves show no respect for national borders.

The international shortwave broadcast bands, as established by the International Telecommunications Union (Table 1-1), are the most popular among radio DXers. Antennas for these and other bands are given special emphasis in Part 1 of this book. Other listeners concentrate on medium wave and/or FM broadcasting (Table 1-2). Additional listeners tune the longwave spectrum as well as the marine and aviation HF bands and additional services often lumped together and called the utility services (UTE).

Table 1-1. ITU International Broadcast Band

Band (Meters)	Frequency Range (kHz)	Band (Meters)	Frequency Range (kHz)
120	2300–2498	25	11650–12050
90	3200–3400	21	13600–13800
75	3900–4000	19	15100–15600
60	4750–5060	16	17550–17900
49	5950–6200	13	21450–21850
41	7100–7300	11	25600–26100
31	9500–9900		

Table 1-2. Broadcast and Longwave Services

Service	Frequency Range
Longwave	20–535 kHz
Medium Wave Broadcasting	535–1605 kHz
FM Broadcast Service	88–108 MHz

The VHF/UHF spectrum between 30 and 900 MHz is the listening territory of the scanner enthusiasts (Table 1-3). In this area there are landmobile, marine, aviation, and public services transmissions along with other radio activities. It has become an ever growing phase of radio listening. Ham radio operations and one-way SWL ham-radio listening are both increasing worldwide, (Table 1-4). Antennas for these activities are subject matter for Part 2 of this book. The advent of the modern all-band receiver with sideband demodulation capability permits radio listeners to enjoy conversation among radio amateurs from around the earth and via satellite.

Table 1-3. VHF/UHF Scanner Frequencies

Band	Frequency Range (MHz)
VHF Lo	30–50
Aviation	118–136
VHF Hi	148–174
VHF Marine	156–162
UHF Lo	450–470
UHF Hi	470–512

Table 1-4. Radio Amateur Bands

Meters	Frequency Range (MHz)	Meters	Frequency Range (MHz)
160	1.8–2.0	12	24.890–24.990
80	3.5–4.0	10	28.000–29.700
40	7.0–7.3	6	50.000–54.000
30	10.100–10.150	2	144.000–148.000
20	14.000–14.350	1¹/4	220.000–225.000
17	18.068–18.168	70 cm	420.000–450.000
15	21.000–21.450		

The frequency bands listed in Tables 1-2 through 1-4 are appropriate for the USA and much of the world. However there are significant differences in various parts of the world for some of the special radio services and their frequencies. The set of constants associated with the dimension tables can be used to calculate antenna lengths for frequencies not included in the tables.

1-4. Time Zones

The earth's surface is subdivided into time zones that correspond to the earth's rotation with relation to the position of the sun. Coordinated Universal Time (UTC) (previously called Greenwich Mean Time—GMT) is the time present on the meridian passing through Greenwich, England. From this starting point, time on the earth's surface is subdivided into 24 15-degree sectors. These zones are numbered positive from one to twelve toward the west of Greenwich and negative from one to twelve toward the east. For example for dawn to arrive in Washington, DC, the earth must spin or advance (+5) zones before the earth is in the position for dawn to occur. Hence, the +5 zone corresponds to the Eastern Standard Time zone (EST) of USA. Thus Washington DC, as an example, is five hours behind UTC time. Therefore, 1200 (noon) UTC time corresponds to 0700 AM Washington DC time. Charts and even dual clocks are available that permit you to determine quickly your local time for any given value of UTC time. Details can be found in *World Radio TV Handbook* and other shortwave log books. Refer to the list at the end of this chapter.

1-5. Antenna Orientation

When a directional antenna (one that displays more sensitivity in one compass direction than others) is to be oriented toward some favorite country or continent it is necessary that you know the bearing angle toward the part of the world to which your antenna is to be made most sensitive. A good compass and a great circle azimuthal map are helpful. Such a map can be used to find direction and shortest distance from one point on the earth to any other point based on our spherical earth. Fortunately the directional antennas covered in this handbook are not sharp-beam types and bearing accuracy is not highly critical. The first step is to determine true north at the antenna site using the polar star, the sun and a vertical pole at noon, or a compass and proper declination correction. An azimuthal map shows true bearings to all parts of the world from one point. Such maps are available from the American Radio Relay League (ARRL). Refer to the listing at the end of this chapter for several sources. You can also purchase such maps from computer services that sometimes advertise in the ham radio magazines. Such special maps position your own city or town in the center to give you accurate readings to all parts of the world.

Some computer services also supply tables based on your location showing true bearings to most of the countries of the world and their principal cities. For shortwave broadcast listeners there are tables that present the bearings in terms of the location of the transmitter sites of the international broadcast stations. A variety of other data such as latitude, longitude, distance, etc., is also given.

1-6. Information Addresses

Sources of information are invaluable in extending your radiocommunications hobby. Reading and learning keep you active and prevent hobby boredom. Addresses of antenna manufacturers and sources are found in the appendix.

Amateur Radio Books and Magazines

American Radio Relay League (QST), Newington, CT 06111

CQ Magazine, 76 North Broadway, Hicksville, NY 11801

Ham Radio Magazine, Greenville, NH

73 Magazine, WGE Center, Peterborough, NH 03458

Radio Clubs and Publications

IRCA, P.O. Box 432, East Moline, IL 61244

Longwave Club of America, 45 Wildflower Rd., Levittown, PA 19057

NASWA, 45 Wildflower Rd., Levittown, PA 19057

NRC, Ron Musco, P.O. Box 118, Poquonock, CT 06064

SPEEDX, 7738 East Hampton St., Tucson, AZ 85715

Scanner Association SCAN, P.O. Box 414, Western Springs, IL 60558

Worldwide TV-FM Association, P.O. Box 514, Buffalo, NY 14205

Radio Broadcast Data and Logging

Radio Database International, P.O. Box 300, Penns Park, PA 18943

World Radio TV Handbook, Century Printing, 6059 Essex, Riverside, CA 92504

Radio Listening Magazines

Popular Communications, 76 N. Broadway, Hicksville, NY 11801

Monitoring Times, P.O. Box 98, Brasstown, NC 28902

Guide Books

Air-Scan Guide, and other publications, CRB Research, P.O. Box 56, Commack, NY 11725

Beacon Guide, Ken Stryker, 6350 N. Hoyne Ave., Chicago, IL 60659

Tiare Publications, P.O. Box 493, Lake Geneva, WI 53147

Great Circle Tables and Maps

Bill Johnston, N5KR, 1808 Pomona Drive, Las Cruces, NM 88001

American Radio Relay League, Newington, CT 06111

Meridian Crossings and Programs

Wm. A. Borchers, 460 Hale St., Suffield, CT 06078

Commodore 64 Antenna Programs

C-64/128 Programs for Amateur Radio and Electronics, by Joe Carr, Howard W. Sams & Co., Indianapolis, IN 46268
(Recommended for its useful antenna, transmission line, and propagation path programs.)

Chapter 2

Basic Antennas and Their Construction

Be Safe

Safety is a prime personal responsibility when erecting and installing antennas. Electric shock, physical injury, and damage to property are dangers. Plan your antenna well and organize your procedures wisely. Be certain to position your antenna and mast where it cannot come in contact with or drop upon power lines when it is being installed or in a storm, or upon a person or neighbor's property when it is being erected or in a storm, or when its parts become fatigued. Use a wire quality and size that can withstand the stress of erection and weather. An added safety factor is provided by insulated antenna wire. A high installation increases the magnitude of all the danger factors of concern. Try to anticipate hazards before you go ahead with the installation.

THIS chapter introduces the basics—basic antenna types and basic constructions. Unlike many technical fields the basic antenna structure is valid, real, and efficient. Each antenna described in this chapter is a good one. Recognize that the number of basic antennas in practical use today far outnumber the working complex and elaborate antenna systems. So this is an important chapter.

2-1. Introduction

The fundamentals introduced are important to the radio amateur and to the radio listener. For simple low-cost and effective antennas you need not go any further than this chapter. But, I have a feeling you will read on as you become interested in something a bit better, more convenient, or something that meets some specific need of your radio room activities.

Most of the antennas can be used immediately for ham radio application as well. Assign a little additional care in trimming them to exact true resonance, proper attachment of a transmission line, and the establishment of a low standing-wave ratio (SWR). All this is clarified in Chapter 12. Dimension information for ham radio use is given in the tables in Chapter 16.

2-2. Basic Antenna Types

There are many varieties of basic antenna styles. However, most antennas stem from the fundamental Hertz and Marconi types shown in Fig. 2-1. These antennas are said to be resonant because their physical lengths are related to the wavelength of the RF energy they must radiate or receive. The Hertz antenna is usually called a half-wavelength or half-wave antenna because its physical length corresponds to that dimension of a wave. Thus the actual physical length of an antenna is related to the frequency of the signal it must receive or transmit.

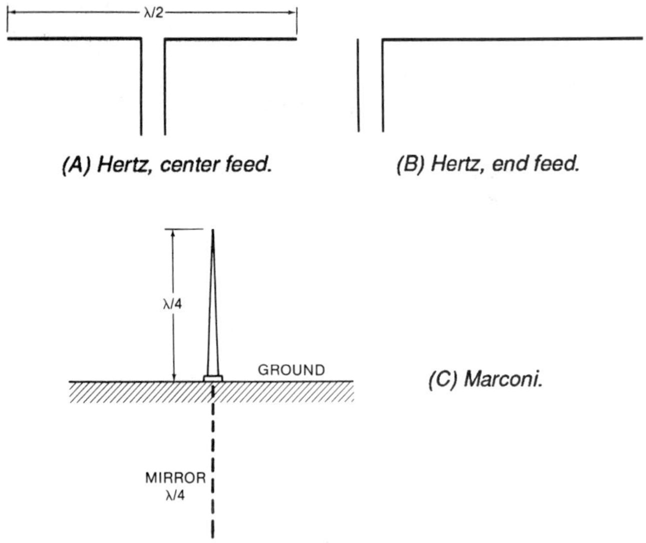

Fig. 2-1. The basic Hertz and Marconi antennas.

Radio-frequency energy is most often supplied or extracted at the center of the Hertz antennas as shown in Fig. 2-1A. Such an antenna is referred to as a half-wave dipole. However, it can be fed at the end as shown in Fig. 2-1B or at some other position along its length.

Physically, a Marconi antenna is one-quarter wavelength long. Electrically, it acts like a half-wavelength antenna because the ground beneath it forms a mirror quarter-wave segment as shown in Fig. 2-1C. The characteristics and quality of the mirror segment are related to ground conditions. A network of conductors (radials) or large-area metallic surface is often laid out beneath the antenna to provide a more uniform or lower resistance ground system.

The quarter-wave element can be mounted on a base insulator or grounded directly. In practice, the former is more common because the results are more uniform and matching is less of a problem.

A horizontal antenna radiates a horizontally polarized wave. This means the electric vector is horizontal with respect to the earth's surface. A vertical antenna radiates a vertically polarized wave with its electric vector vertical or perpendicular to the earth's surface. If a vertically polarized wave is desired, it is possible to mount a half-wave dipole vertically.

A very practical formula for calculating a half wavelength is:

$$\lambda/2 = 492/f$$

where,

$\lambda/2$ is the length of a half-wavelength in feet,
f is the frequency in megahertz.

The free-space dimension calculated from this formula represents the half-wavelength of a radio wave travelling in free space. Because a radio wave travels slower in most other media than in air, the half-wavelength is shorter than the free-space calculation. This means a radio wave travels slower in a transmission line or antenna than in free space; therefore, the separation between adjacent waves is less than in free space.

A similar relationship exists for a radio wave on an antenna; a half-wave antenna is physically from 1- to 6-percent shorter than its free space measurement. A point of discontinuity between the antenna conductor and space exists at the end of the antenna. To a degree, space acts as a dielectric material and the capacity effect links the two ends of the antenna. This end effect lowers the resonant frequency of the antenna. Antenna diameter also has a very pronounced effect on exact resonant length. For this reason it must be cut shorter than the free-space calculation to obtain resonance at a given frequency. In other words, an antenna must be cut to an electrical half wavelength which is physically shorter than a free-space half wavelength. This factor is considered in the extensive antenna dimension tables that are a part of this book.

A thin antenna element must be cut only slightly shorter than its free-space value, whereas a large-diameter antenna element must be made significantly shorter. Since a shorter physical length is required at the higher frequencies, the length is much more critical at these frequencies.

The formula most often used for calculating antenna resonant length for calculation in the high-frequency spectrum is:

$$\lambda/2 = 468/f$$

where,

$\lambda/2$ is the length of a half-wavelength antenna in feet,

f is the frequency in megahertz.

A $\lambda/2$ expressed in meters for the previous two equations would be:

$$\lambda/2 = 150/f$$

$$\lambda/2 = 142.5/f$$

When cutting a dipole antenna each segment is one-half of this value or the equation becomes:

$$\lambda/4 = 234/f$$

where,

$\lambda/4$ is the length of a quarter wave in feet,

f is the frequency in megahertz.

and

$$\lambda/4 = 71.25/f$$

where,

$\lambda/4$ is the length of a quarter wave in meters,

f is the frequency in megahertz.

2-2-1. Dipole Types

The three most common dipole types are shown in Fig. 2-2. Additional dipole variations are covered in Chapter 5. The straight and conventional horizontal dipole of Fig. 2-2A is usually suspended between two masts. If it is an especially long dipole an additional center mast may be advisable to prevent sagging.

The inverted-V antenna of Fig. 2-2B is a popular style. In this arrangement only a single high mast is required. The two ends of the antenna can be supported by short posts. This is a very popular style of dipole because it is more omnidirectional than the straight dipole. Of course it is more economical to erect. Also, the dipole ends are more readily available and it is more convenient for trimming and making other changes. There is also some limited low-angle sensitivity off the antenna ends, 90° related to the normal broad figure-eight pattern of a dipole.

The sloper configuration of Fig. 2-2C shows one end of the antenna up and the other end down. This arrangement produces some additional directivity in the direction of the downward slope as indicated by the arrow. Apparently there is some limited additional gain in this direction and less gain at the complementary direction. However it can be considered reasonably omnidirectional.

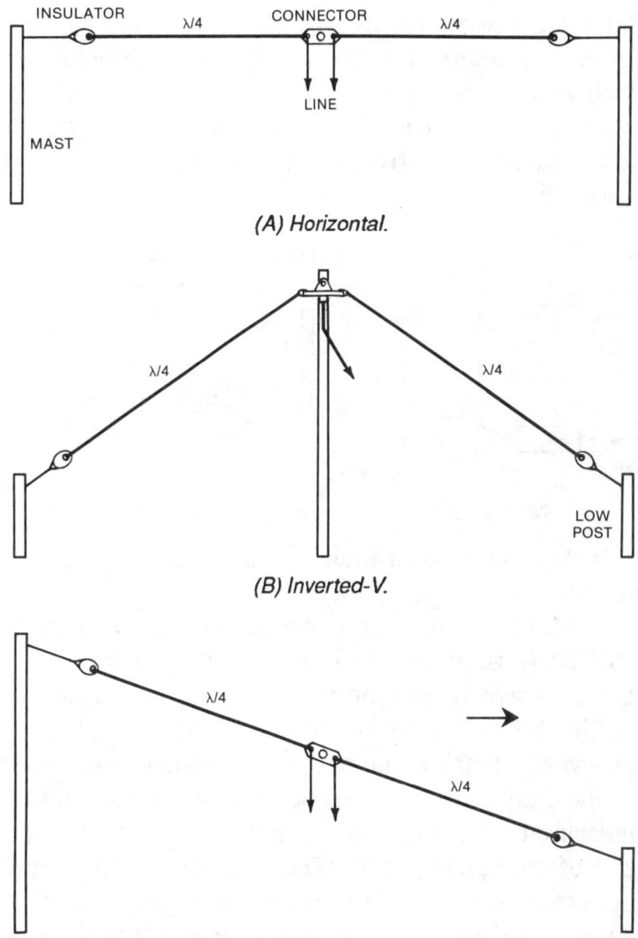

Fig. 2-2. Three basic dipole types.

The three basic antenna styles of Fig. 2-2 are perhaps the most widely used antennas by radio listeners and ham operators. If you add to this the quarter-wave versions of Fig. 2-3 the group will outnumber by far all other antenna arrangements. Actually they are an integral part of most of the antennas to be described in this handbook.

2-2-2. Quarter-Wave Antenna with Ground

The quarter-wave vertical antenna of Fig. 2-3A is widely used for reception and transmission in the

VHF/UHF spectrum as well as for the mid- to high-frequency segment of the HF spectrum. It does very well as a receiving antenna for the higher-frequency shortwave broadcast bands. Special versions of this antenna for low-frequency use are presented in Chapter 4.

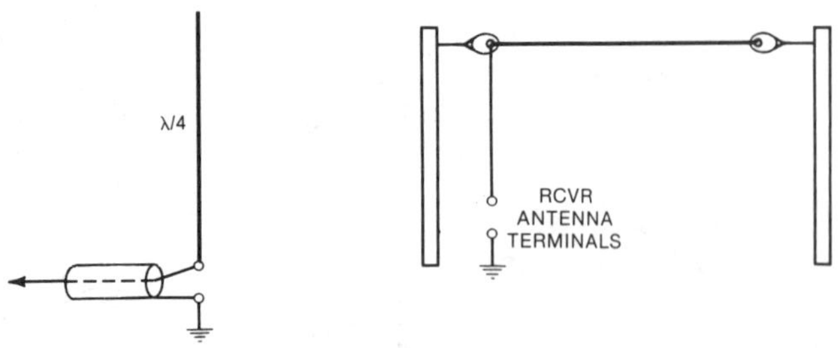

(A) λ/4 vertical. *(B) λ/4 single wire.*

Fig. 2-3. Quarter-wave antennas with ground.

Many of the common shortwave broadcast (SWB) horizontal receiving antennas are actually quarter-wave wires working against ground as shown in Fig. 2-3B. Usually, in the case of Fig. 2-3B, the single-wire antenna is quarter-wave resonant somewhere in the frequency range to be received. This length is measured from the receiver antenna terminal to the end of the antenna wire. The receiver and its ground system acts as a ground for the antenna system. In some portion of the frequency spectrum to be received, it acts as a quarter-wavelength antenna and presents an antenna resistance that matches the low impedance antenna terminal of the receiver. At other frequencies it is less than favorable because of rising antenna reactance and impedance. There are exceptions, because at some much higher frequency it becomes some odd multiple of a quarter wavelength.

The quarter-wave single-wire antenna is often referred to as a longwire. More correctly though a longwire antenna is one that is greater in length than one or two wavelengths. This certainly is not so for most of the antennas used for the medium wave (MW) and tropical band reception.

Why is the longwire length so significant? The free end of the antenna has a high impedance because it is the high-voltage and low-current point of the antenna system. Exactly one quarter-wavelength from this point the wave distribution of the antenna is such that there is a maximum current and a minimum voltage and, therefore, a low impedance. Thus at a point one-quarter wavelength from the end, the antenna matches the low impedance input of the receiver and maximum current is transferred to the receiver input.

When a transmission line that connects between the antenna and receiver is matched at the antenna and matched again at the receiver these impedance changes do not occur or are very limited. The practical significance of this is that no matter how long the transmission line is made the same relationship exists. If there is a low impedance match at the antenna, a low impedance appears at the other end of the transmission line regardless of the overall length of that line. As a result there is a consequent low-impedance match to the low impedance input of the receiver no matter what length the transmission line happens to be. Again the antenna system supplies maximum signal current to the receiver input. More about this subject next.

2-3. The Transmission Line

An antenna system is much more than the antenna wire or tubing. There is the transmission line that must be connected to the antenna which conveys the signal to the receiver or, in the case of ham radio operation, also transfers the transmitter output to the antenna.

The most popular transmission line used for radio listening and, for ham radio operation at power levels of no more than 200 watts on frequencies up to about 60 MHz are 50-ohm and 70-ohm coaxial line (RG-58/U and RG-59/U respectively) and 300-ohm flat line. Another coaxial line used widely with scanners and other VHF/UHF equipment is RG-6/U because of its low loss and long life expectancy. A good quality RG-58/U or RG-59/U line does very well over most of the HF bands especially for lengths that do not exceed 150 ft (45.7 m). At the high-frequency end of the HF band and into the VHF/UHF spectra they are acceptable for lengths up to 100 ft (30.5 m) in most applications. RG-8/U is a larger-diameter 50-ohm line with lower losses over the HF and VHF spectra.

More expensive coaxial cables are available. For example, the types listed before with a suffix A/U (as in RG-59A/U) have a longer life span. Cables with low VHF and UHF losses are RG-6/U and RG-58C/U and the larger diameter RG-8F foam line. In my own situation with a line length usually well under 150 ft, I have used quite ordinary RG-58/U and RG-59/U with good results and no lack of signals and contacts over many years. And I hear ham stations pounding in from all over the world and from lands where expensive lines are not available or not affordable. You can continue to enjoy radio listening and ham operations on a low budget.

2-4. Easy-Up Masts and Supports

A major objective of this book is to provide information about good-performing low-cost antennas. Cost considerations also involve the antenna mast. Costs and responsibility rise tremendously when you begin to erect masts that must stand much more than 30 ft (9.144 m) above ground. This is not to say that such antennas do not pick up longer distance signals somewhat better with their greater elevation. However, the economic aspects must be considered and sometimes it isn't worth the additional cost, aggravation, legal responsibility, and replacement hassle if the antenna doesn't stand up in the wind. We've been told often that if we have a 30-ft (9.144-m) high antenna in the clear, the additional improvement is only significant if one takes a jump to 60 ft (18.29 m) especially so for medium-wave and low-frequency HF operation. The story is a bit different at the high-frequency end of the spectrum where beam antennas on rotors can be used. However, such antennas are not in the realm of limited finances. All you need do is price a typical high mast, antenna rotor, and beam antenna trio.

Height is more of a factor in VHF/UHF operation because of the extended line-of-sight propagation characteristics. However you can still do a lot with an antenna height of 30 ft (9.144 m) and either a homemade or purchased low-cost beam. The physical size of a beam in this spectra is small. Thus the 30-ft (9.144-m) height and a high gain beam does an admirable job.

My own personal experience may give you some encouragement. Seldom did I have a mast higher than 30 foot (9.144 m) and some as low or lower than 20 foot (6.1 m). In my years as a radio amateur I had pleasurable contacts with all continents and have worked 100 countries more than once. In fact, a properly oriented gain antenna at low level often does better than a simple antenna erected 50 or more percent higher. Unless otherwise indicated the masts used for the antennas constructed for this book were 20 ft (6.1 m) to 25 ft (7.62 m) tall.

Propagation conditions are the great signal leveler. If propagation favors the transfer of a shortwave signal between two locations the signal exchange can be infinitely better between the station and your low antenna than between the same station and another receiving location perhaps only 100 or so miles from you. There may be at the moment a poor propagation path to the second receive location despite the availability of a significantly higher antenna.

Fig. 2-4. A PVC mast in position.

The antennas described in this book were erected and tested using economy masts such as TV telescoping antenna masts or masts constructed of PVC piping sections telescoped together and bolted (Fig. 2-4). Since the appearance of PVC plastic piping it has become our favorite for lightweight mast construction. Such a mast is lightweight and easy to handle. It is an insulator and a variety of antenna styles and physical arrangements can take advantage of this insulating quality. Such a mast (Fig. 2-4) can take the weather well. A low mast need not be guyed or require just a single guy used to hold it exactly vertical. For higher versions or when the antenna puts a considerable strain on the mast you can use three or four simple guys.

2-4-1. PVC Mast Erection

A mast somewhat longer than 18.5 feet (5.64 m) can be constructed by telescoping a 10 ft (3.1 m) 1.5-in ID section of PVC plastic tubing about 15 inches into a 10-foot section having a diameter of 2 inches (5.1 cm) ID as shown in Fig. 2-5. Two $1/4$-in (0.64-cm) diameter, 3-in (7.62-cm) long bolts and associated lock washer and nut hold the pipes together. The light weight of this mast surprises you and you are able to lift it up and down with ease. It can be held erect by driving a metal fence post into the ground as shown in Fig. 2-6A. Drive the post into the ground as far as the top of its triangle. Mount a through-bolt in the mast about 15–18 inches (38.1–45.7 cm) from the bottom. You can now easily lift the mast over the top of the metal fence post and let it drop over the post as shown in Fig. 2-6B. The through-bolt will rest on top of the fence post and you will have added a foot or two (depending on the length of the fence post) to the height of the mast taking the height of the mast up to the 20-ft (6.1-m) level.

Fig. 2-5. Top section telescoped into bottom section and fastened with two through-bolts.

You have a versatile mast. You can lift it up and down with ease to make changes or to try different antennas. You can move it to another position simply by moving the metal fence post or driving a second metal fence post into the ground at a different position.

An alternative plan is to use a round metal fence pole or piece of pipe (Fig. 2-7). Such an arrangement will provide a more permanent installation. The pole can be of any practical length that you desire. Excavate a hole with a post hole digger. Go below the freeze line as far as practical and support the pole with stone or, better, with some ready-mix cement. Again use a through-bolt to hold the mast above ground level to obtain additional height.

PVC piping also is available in 20-ft (6.1-m) lengths if you have suitable transportation between the source and your home or if you can get delivery. This would give you an additional foot or more of mast height because it need not be telescoped.

Three sections of PVC piping can be telescoped. In this arrangement an additional 6-ft (1.83-m) section of 1-in (1.5-cm) ID piping is telescoped into the top of the $1\frac{1}{2}$-inch (3.8-cm) ID piping to add an additional 5 ft (1.52 m) of length to the mast. This mast you can handle yourself and lift it over the fence post or pole. You have much less trouble when there is no wind. Figure 2-8 shows such a mast erected just outside the radio room. A 19-meter vertical has been mounted on the mast. This antenna serves as a reference antenna when checking out other 19-meter antennas. 19-meters is a favorite band and the size of the antennas used for this band are relatively short making it convenient to test out various antenna types at this frequency before scaling them up or down to operate on other shortwave bands.

In this protected location I use a pair of guy

(A) Metal fence post driven into ground.

(B) Mast sitting over top of post.

Fig. 2-6. Mounting a mast over a metal fence post.

wires to hold the vertical erect and to keep it from flexing too far away from the arrival direction for prevailing winds. The support for the guys on the mast is a 4-in (10-cm) eye-ring bolt. This bolt is used instead of a regular bolt in one of the positions where the top telescoping pair are bolted together.

Telescoping TV masts also serve as an economical means of erecting wire antennas. Radio Shack has two standard lengths of 19 ft (5.79 m) and 36 ft (10.97 m). One person can handle the shorter one with just a little more difficulty than the lighter PVC piping masts of the same height. The 36-ft model comes under the erection category of "difficult-for-one but easy for two." Work with shorter masts first before you tackle the latter one if you have not had previous experience.

2-5. Construction of a Quarter-Wave Vertical

In the next five topics the construction procedures for each of the five basic antenna types (quarter-wave vertical, dipole, sloper, inverted-V, and random wire

(A) Metal fence pole set in ground. *(B) PVC mast sitting over top.*

Fig. 2-7. Mounting a mast over a round metal fence pole or pipe.

[longwire] are detailed). There are two important reasons for you to go over each description carefully. First, they represent by far the most widely used antennas for both reception and transmission and, second, you are most likely to erect one or more of these types before you advance to antennas that are more complex or to some modification or grouping of types. Consequently it is important to be completely familiar with the basic types from which they evolved. All the basic types are good performing antennas and you can anticipate fine results for either reception or transmission.

A second advantage of paying particular attention to these early topics has to do with the assembly and erection details given and, other practical procedures of antenna installation. Much of this information applies to all antennas but will not be repeated in the same detail in the chapters that follow. In fact in constructing some of the later antennas you may refer to this section quite often.

The quarter-wave vertical antenna is a good performer and does not require much space. Over an adequate ground it will perform as well or better than longer types and it is especially effective on the higher frequency shortwave broadcast bands. Its application on the lower frequency bands is restricted by the

Fig. 2-8. PVC mast erected near a house.

height of the structure needed to support its length. For example a quarter wave in length on the 49-meter band is approximately 38 ft (11.6 m) and it is considerably longer for the 60-, 75-, 90-, and 120-meter bands as the dimension tables indicate. However, in Chapter 4 you will learn how the basic quarter-wave vertical can be modified to obtain good performance on the lower-frequency bands along with reduced overall height.

Construction details for a 19-meter quarter-wave vertical attached to a PVC mast are shown in Fig. 2-9. The length of the vertical wire should be 15-ft, 3-in (4.65-m) as shown in the 19-meter segment of the antenna dimension Table 11-1. Wire used in the construction of this antenna and all of the listener and radio amateur wire antennas of this book was vinyl-covered No. 14 AWG (approximately 1.6-mm in diameter) stranded. There is no problem for this wire to handle 200-W PEP in the amateur services.

First assemble the mast by dropping the top pipe 16 inches (40.6 cm) into the bottom one and pull up tight on the two through-bolts that hold the telescoped sections together.

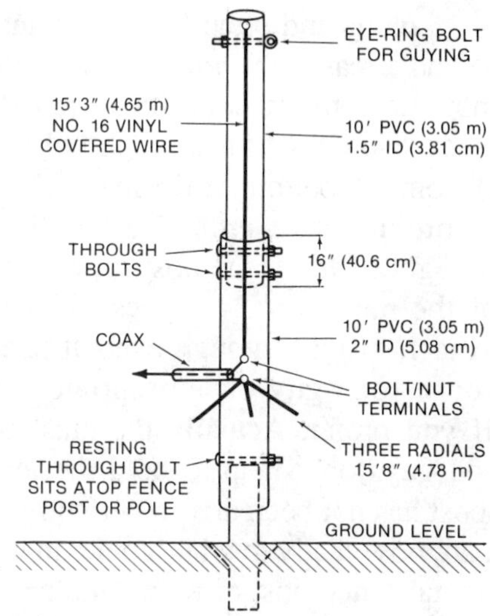

Fig. 2-9. A 19-meter vertical antenna mounted on PVC piping mast.

The vertical wire is cut to length and an eye-ring solder lug is soldered to each end. One end is fastened to the top of the mast with a short bolt and nut combination (Fig. 2-10). The antenna wire is stretched out taut and a dot is penned on the mast to mark the drill point for mounting the top bolt/nut coax terminal. A second hole is drilled about 1.5 in (3.81 cm) beneath

Fig. 2-10. Vertical fastened to top of mast.

for attachment of three radials. Radial length is also shown in the dimension table and is listed as 14 ft 8 in (4.78 m) for this design.

The coaxial line eye-rings or spade lugs connect at the rear between two nuts placed on each of the terminal bolts. The antenna resistance of a quarter-wave vertical is 36 ohms and provides a good match to 50-ohm cable and a reasonable match to 70-ohm cable.

Do not attach the coaxial line or radials until the mast is erect. Mount a fifth through-bolt about 18 in (45.7 cm) from the bottom of the mast. This will raise the mast bottom above ground a length dependent upon the height of the fence post. In the example the bottom of the mast was 34 inches (86.4 cm) above ground. Attach a sixth through eyebolt near the mast top. Tie on a rope guy of appropriate length. Add another if you prefer. Actually the mast is self-supporting and guys only help to hold the mast straight up if the post has not been driven perfectly straight.

You are now ready to walk up the mast and lift it over the metal fence post or pole. One person can do it with ease. The top of the vertical will be about 20 ft (6.1 m) above ground.

If you wish about 5 ft (1.524 m) additional height you can construct your mast of three sections of PVC piping. Telescope a 6-ft (1.83-m) section of 1-in (2.54-cm) ID pipe 1 ft (0.31 m) into the top section. A bit more struggle is needed to set the mast over the post, especially on a windy day. However, it is no big problem. Ask for a little help with the first one if you are concerned.

After the mast is in position solder the ends of the three radials to one eye-ring solder lug. On a stepladder you can fasten the radials to the antenna's ground terminal as shown in Fig. 2-11. Stretch out the radials and hold them down with short stakes driven into the ground (Fig. 2-12A). Space them 120° around the mast. Attach the transmission line as shown in Fig. 2-13 and you are ready to DX on 19 meters.

Fig. 2-11. Attachment of the three radials.

If you are cramped for space or wish to eliminate the physical hazard of the low radials drop the radials-directly to the ground as in Fig. 2-12B. Bury them 1 to 2 inches (1.54 to 2.08 cm) beneath the ground; the results will be the same. Even the simple plan of Fig. 2-12C works out well, giving you a simple L arrangement using a single radial. My own permanent installation of Fig. 2-8 does just that because I can hide the radial in the limbs of the nearby evergreen trees. This is my 19-meter reference antenna. It is a jewel on this band and does reasonably well on the adjacent 21- and 16-meter bands.

More than one vertical can be mounted on the same mast as shown in Fig. 2-14. Use separate trans-

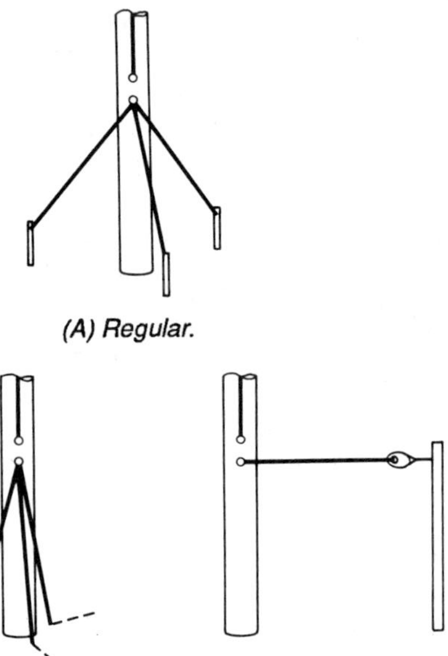

(A) Regular.

(B) Space saving underground. *(C) Single radial.*

Fig. 2-12. Various radial arrangements.

Fig. 2-13. Attachment point of the coaxial transmission line.

mission lines to link the three verticals with a three-position coaxial switch in the radio room. For some time I have operated such a two-band 19- and 25-meter combination with good results.

In the discussion of antennas that follow in later chapters the emphasis will concentrate mainly on the antennas; the mast construction and other mechanical procedures are much abbreviated.

2-5-1. Weatherproofing

When a permanent installation is to be made the hardware can be weatherproofed using sealant tape and silicon rubber sealant. By so doing you prevent hardware rust and ensure a low resistance radio-frequency connection. The example in Fig. 2-15A shows two antenna terminals without sealer while Fig. 2-15B shows the application of silicon sealer. Figure 2-15C shows the application of sealer at the point where the coaxial line end lugs are connected to the two antenna terminals. Also the wire ends have been wrapped with sealant tape. Coaxial and other fittings can be sealed to minimize connector deterioration in the weather. The examples of Fig. 2-16 show such a junction that has been wrapped with sealant tape. In my own situation, antennas are changed frequently and I use sealer only when an antenna is to be operated over a long period of time. The antenna photographs you see in the later chapters will not show such sealants.

2-5-2. Lightning Protection

The chances of a lightning strike are very slim especially if the antennas are low. I have never had a strike despite extensive antenna work over a long number of years. That does not mean I take the subject lightly. There is always that possibility. *Take all precautions.*

There are lightning arrestors and discharge devices for connection to your incoming coaxial line. Only to a degree can they fend off strikes. Certainly they are better than nothing. You can also set up external ground and switching arrangements that make the necessary disconnects and ground the antennas. When you have so many lines coming in, then you must do better. In my situation I find it more convenient and effective to keep the lines exiting the radio room under 20 ft (6.1 m). Externally connectors are used to tie in the antennas to be used or evaluated. When

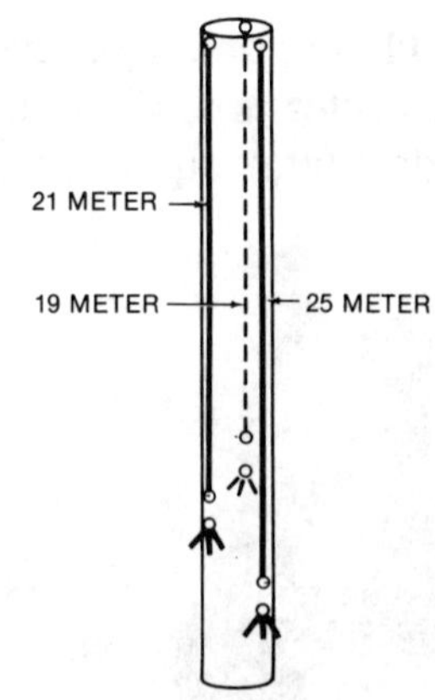

Fig. 2-14. Mounting two or three verticals on same PVC mast.

storms are in the area, are predicted, I have retired, or am away from the house completely, disconnects are made during the stormy months of the year.

Remember that the longer the antenna, the more vulnerable it is to a strike. Banana plugs can be used to break single wire or flat lines. Female-to-female banana connectors are available, if you can find them. A good substitute is the coaxial female-to-female connector (PL-258) in Fig. 2-17.

In general, radio room equipment is more likely to be damaged by the more frequent lightning strikes on the power lines. Do you have a master switch that disconnects all your equipment? Is all your equipment grounded well? Pulling plugs is always a good idea during the thunderstorm months. When a storm is in the area or predicted this procedure provides a more complete break than a switch. Don't expect a lightning strike but don't ignore the possibility either.

2-6. Construction of a Dipole

The horizontal dipole and the inverted version are probably the most common antenna of all. This design performs very well on the band for which it is cut and also serves as a universal standard for checking out the performance of other antennas. Results are also good on the wire's third harmonic frequency (approximately three times the cut frequency) and higher odd harmonic frequencies as well. As a receiving antenna good results are often obtained on the next highest frequency band and sometimes on the adjacent lower frequency band if there is not too great a frequency separation. At odd harmonic frequencies the antenna resistance at the center is low and only a bit higher than at the fundamental dipole frequency. Thus it

(A) Bare connector.

(B) Connector wrapped with sealant tape.

Fig. 2-16. Weatherproofing a coaxial splice connector.

matches a low impedance transmission line which makes it ideal for both reception and transmission. Such specialized antennas are covered in Chapters 5 and 7.

A dipole consists of two equal-length quarter-wave segments. Hence you will find a required seg-

Fig. 2-17. A coaxial connector used to join two banana plugs in a single-wire antenna system.

(A) Bare terminals.

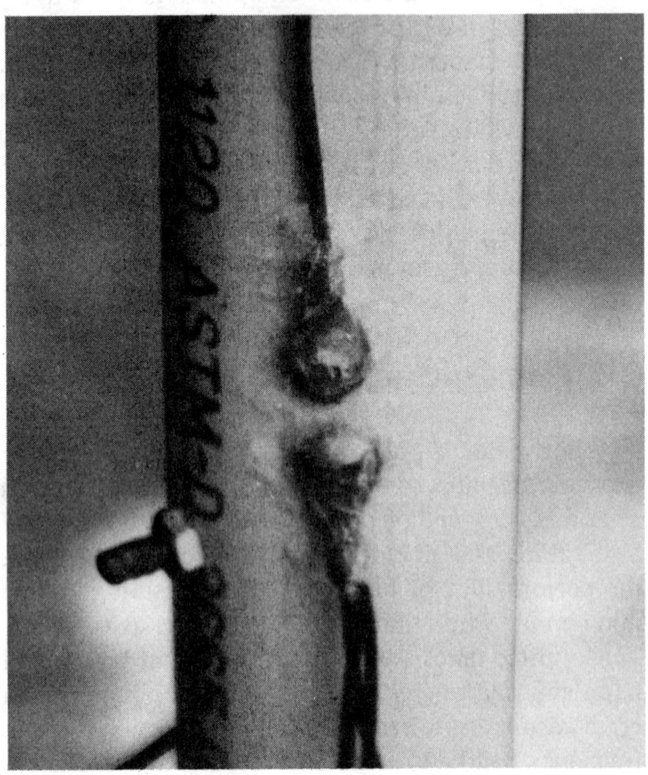

(B) Silicon sealant added to terminals.

(C) Sealant tape and compound added to coaxial line.

Fig. 2-15. Weatherproofing terminals.

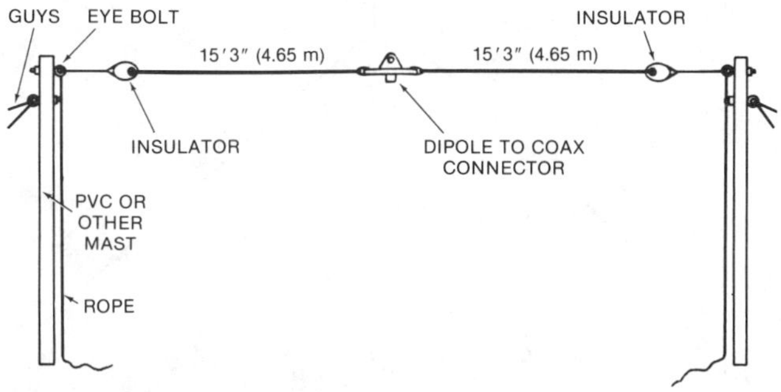

Fig. 2-18. The 19-meter dipole antenna.

ment length in the quarter-wave element column of the dimension tables. This is the same column you use to find the lengths of a quarter-wave vertical for Topic 2-5. In this case you require two such segments of 15 ft 3 in (4.65 m) to construct a dipole.

A basic dipole construction of Fig. 2-18 comprises two masts, two antenna segments, some manner of connecting segment ends to transmission line, three insulators, and preferably some means of raising and lowering the antenna wire without taking down the mast. For a very long dipole you may wish to use a center mast to reduce wire sag.

There are two end insulators that isolate the antenna ends from the support ropes. The insulator at the center can be a dipole-to-coax adapter, affording a convenient means of attaching a line to a dipole using a coaxial plug. An alternative is to use the same type of insulator at the center as you do at the ends. Here the center conductor of the coaxial transmission lines connects to the antenna on one side of the insulator and the braid connects to the wire on the other side. For a permanent connection the wires should be soldered and wrapped with a sealant tape.

For a dipole connector the antenna wire is fed through a hole in the insulator end and then twisted along the insulator lead as shown in Fig. 2-19A; solder for a permanent installation. Again use a sealant tape (Fig. 2-19B) if the antenna is to be up for a long time.

A guy rope can be fed through the eye of an eyebolt attached to the top of each mast before the masts are erected. Perhaps your masts are already in place with a lanyard attached. Each mast can usually be supported with a pair of top guys, spaced 120° on each side of the antenna wire. A third guy does give some extra support if the antenna wire breaks.

(A) Bare connector.

(B) Connector wrapped with sealant tape.

Fig. 2-19. The dipole connector and weatherproofing.

The theoretical dipole feedpoint resistance is 72 ohms but it varies as the function of height above ground. You can use either 50- or 70-ohm coaxial cable. Good results are also obtained with one of the flat lines up to about 300 ohms. Flat lines have even less attenuation and can be of benefit when the transmission line must be longer than 100 feet (30.5 m).

2-7. Half-Wave Sloper for 19 Meters

The sloper is a dipole hung in a different manner (Fig. 2-20). Note that one advantage of the sloper is that it requires but a single high mast. The other end of the antenna is near ground level and can be held down by a metal fence post or pole. The distance from ground at the end of the antenna wire should be 5–7 ft (1.52–2.74 m). Again two guy ropes near the top and spaced 120° on each side of the antenna wire can provide support for the single high mast.

The sloper has become a popular version of the dipole over the last two decades. It is a good performer and displays limited directivity. The characteristic figure-eight pattern of the dipole appears to be tilted forward in the direction of the antenna wire tilt as shown in Fig. 2-20. Long distance DXing is good

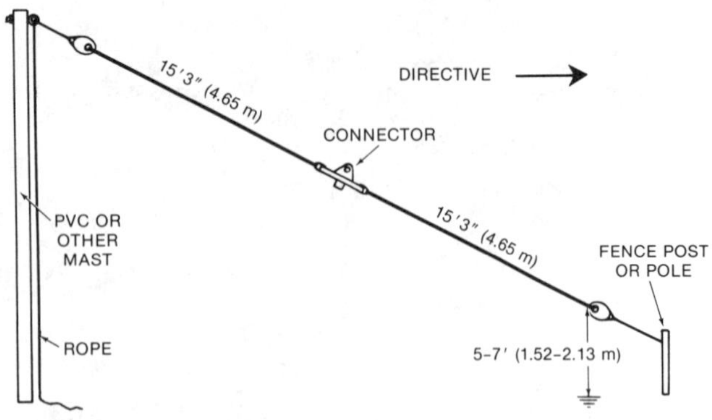

Fig. 2-20. The 19-meter sloper antenna.

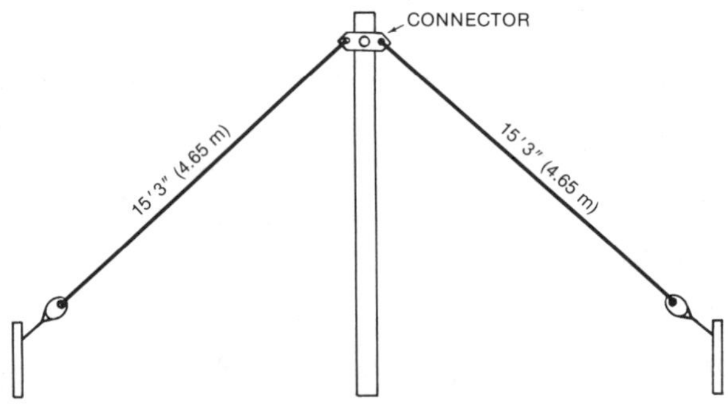

Fig. 2-21. The inverted dipole antenna.

with the sloper in this direction indicating some good low-angle pickup. Back pickup away from the direction of the wire tilt is poorer suggesting there is gain in the tilt direction perhaps only at low vertical angles.

The sloper has another attractive attribute in that one end is low to the ground. Consequently it is easy to change the slope by driving one or more additional posts into the ground at other angles. When you wish to change directivity you free the sloper at its low end, move it, and point it in another direction.

In the erection of a sloper you can let the coaxial cable drop straight away from the dipole-to-coaxial connector. Then in a long sag bring it back to the high mast and tape it. Don't pull it up tight because it is good practice to let the line drop away from the antenna at a right angle for a reasonable distance. It is an inexpensive and easy-to-erect dipole and works well.

Fig. 2-22. Tie-down for an inverted dipole or sloper ends.

2-8. The Inverted-V Dipole

The inverted-V dipole or inverted dipole half-wave antenna (Fig. 2-21) is another mounting modification of the basic dipole. Over a period of time it has supplanted to a degree the strictly straight conventional dipole. Wire ends are brought down to low level and can be held taut with ropes attached to fence posts as shown in Fig. 2-22. It is a very effective inexpensive antenna, and requires but a single mast. The pattern of the inverted dipole is more omnidirectional than a conventional dipole and it shows reasonable low angle pickup even in the direction off its wire ends.

A favorable attribute of the inverted dipole is that both ends are accessible from the ground, therefore, it is easy to trim the antenna wires to obtain some precise resonant frequency. Also wire can be added to both ends very easily if you wish to operate the antenna on more than one band. This add-on subject is covered in Chapter 3.

The length of the dipole element in the inverted configuration is influenced by the height above ground and the angle between the two wires. As the angle becomes more acute the lengths of the two dipole elements must be increased to maintain resonance on a given frequency. The length usually falls somewhere between the figures given in the dimension tables for a $\lambda/4$ element and a $\lambda/4$ radial. It is a

Fig. 2-23. Terminals for antenna wires and coaxial line mounted atop the top PVC mast section.

good idea to cut the lengths for an inverted dipole to those given under the quarter-wave radial column, and then trim them back to reach the desired resonant frequency. For reception the length is not a critical factor. However, if the antenna is to be used for transmission, length could be of concern in setting operating bandwidth. A very low dipole often requires significant trimming when operated on the lower HF frequencies.

The same basic dipole assembly and dipole-to-coaxial line connector arrangement described previously for a conventional dipole and sloper can also be used for the inverted dipole. A lanyard arrangement can be incorporated to pull the connector up and down the single mast. Thus it is possible to change over to another band by switching wire lengths.

If a PVC mast is used you can incorporate built-in antenna terminals by drilling two holes at the very top of the mast and attaching bolt/nut pairs as shown in Fig. 2-23. The end of a 20-ft (6.1 m) length of coaxial line is passed through a hole cut in the 1-in ID (2.54-cm) pipe which is the top of a three-section PVC mast. Solder rings are now fastened to the dressed ends of the coaxial line. Terminal bolts hold these two rings inside the tubing. The terminal assembly is held tight by an external nut and lock washer. Another nut on each terminal will hold the antenna wires which also have been soldered to a solder ring. Plastic caps are available to slip over the mast top, providing weather protection.

2-9. End-Fed Random Single Wire

The random single-wire antenna of Fig. 2-24 is a common shortwave listening antenna. As mentioned previously it is often referred to *very* loosely as a longwire. A common length is 75 ft (22.9 m). By trimming to about 71 ft (22.64 m) a more uniform performance over all the shortwave bands can be obtained. The length represents the distance from the far end of the antenna wire to the antenna terminals of the receiver. A good ground is also important to this averaging process. The receiver input characteristics also have an influence on the length of a random wire that produces good average results over a wide spectrum.

A length of 71 ft (21.64 m) represents a quarter wavelength on the 90-meter band and it works in conjunction with the ground you attach to your receiver or the radio-frequency ground your receiver finds by being connected to the power lines. Recall at an electrical quarter wavelength from the free end of a wire there is a low impedance that results in a reasonable match to the receiver input and transfers an appropriate signal current to the receiver. This good transfer characteristic extends over a reasonable bandwidth.

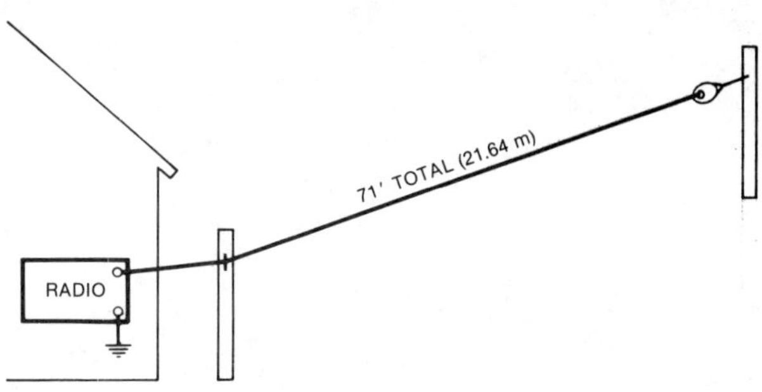

Fig. 2-24. Random-wire antenna.

(Bandwidth is the frequency spectrum over which an antenna will display an essentially uniform sensitivity. For example, an antenna that does well 100 kHz on each side of 3.3 MHz has a greater bandwidth than one which covers a span of only 50 kHz.) However on frequencies away from resonance the match is less efficient and the pickup is not as good as from a longwire that is cut for these frequencies.

A random-wire antenna of this sort, as a receiving antenna, has a broader bandwidth than a dipole cut to the same frequency. However the dipole delivers a better signal than a random wire at its design frequency.

It is important to realize that a low impedance appears at the end of the wire for each odd-multiple of a quarter wavelength. In terms of the international shortwave broadcast bands a third harmonic relation for a random wire 70 ft (21.34 m) in length occurs on the 31-meter band and the fifth harmonic somewhere between the 16- and 19-meter bands. It is these relationships that contribute to the acceptable average performance of a random-wire antenna.

One might jump to the conclusion that a tuner should be a great help in peaking signals on frequencies beyond the bandwidth set by the odd harmonic antenna resonances. Actually improvement is often quite limited because the amount of signal intercepted by the antenna itself off its resonant bandwidth is less. However a tuner sometimes does permit the receiver to take best advantage of this weaker signal received on the lower frequency shortwave bands.

Bookmark

HOWARD W. SAMS & COMPANY

DEAR VALUED CUSTOMER:

Howard W. Sams & Company is dedicated to bringing you timely and authoritative books for your personal and professional library. Our goal is to provide you with excellent technical books written by the most qualified authors. You can assist us in this endeavor by checking the box next to your particular areas of interest.

We appreciate your comments and will use the information to provide you with a more comprehensive selection of titles.

Thank you,

Vice President, Book Publishing
Howard W. Sams & Company

COMPUTER TITLES:

Hardware
☐ Apple 140 ☐ Macintosh 101
☐ Commodore 110
☐ IBM & Compatibles 114

Business Applications
☐ Word Processing J01
☐ Data Base J04
☐ Spreadsheets J02

Operating Systems
☐ MS-DOS K05 ☐ OS/2 K10
☐ CP/M K01 ☐ UNIX K03

Programming Languages
☐ C L03 ☐ Pascal L05
☐ Prolog L12 ☐ Assembly L01
☐ BASIC L02 ☐ HyperTalk L14

Troubleshooting & Repair
☐ Computers S05
☐ Peripherals S10

Other
☐ Communications/Networking M03
☐ AI/Expert Systems T18

ELECTRONICS TITLES:
☐ Amateur Radio T01
☐ Audio T03
☐ Basic Electronics T20
☐ Basic Electricity T21
☐ Electronics Design T12
☐ Electronics Projects T04
☐ Satellites T09

☐ Instrumentation T05
☐ Digital Electronics T11

Troubleshooting & Repair
☐ Audio S11 ☐ Television S04
☐ VCR S01 ☐ Compact Disc S02
☐ Automotive S06
☐ Microwave Oven S03

Other interests or comments: ______________________

Name ______________________
Title ______________________
Company ______________________
Address ______________________
City ______________________
State/Zip ______________________
Daytime Telephone No. ______________________

A Division of Macmillan, Inc.
4300 West 62nd Street
Indianapolis, Indiana 46268

22495

Chapter 3

Antenna Tests and Comparisons

MANY variables influence the performance of an antenna in the field including the site itself. It is a difficult assignment to computerize the performance of an HF antenna and take all contributing factors into consideration. A number of these factors are ground conditions at the site, elevation and slope of the terrain, surrounding metallic objects, and height above ground. In addition, antenna factors such as gain, orientation, and polarization must be considered. Add to these the variations of propagation, vertical angle of signal arrival, and polarization and you have quite a pot of soup to sample!

3-1. Perspective

In a practical sense the best antenna sampling is done by comparison. It is a direct way of judging performance using the world as your signal generator. Such a procedure for the radio listener and radio amateur is interesting and not too costly. Supplies are readily available and reasonable.

Then again, you may not wish to do any testing of antennas and would prefer to choose one or several of the antennas in this book that best meet your needs and available space. This is fine. Just about every antenna in Part 1 of this book has been checked out by comparison using the procedures covered in this chapter. The same applies to the ham antennas of Part 2. The general performance on individual or several bands is mentioned for each one.

3-2. Basic Setups

Propagation conditions change so rapidly that it is difficult to compare any two reasonable antennas by replacing one with the other or by running out in the back yard and changing the transmission line over from one to the other. You must have some means of making a quick antenna changeover. A two or three position coaxial switch permits one to do the job quickly and at little cost for coaxial or single-wire transmission lines (Fig. 3-1). When using a single-wire line attach a banana plug to its end. Such a plug can be inserted into the center connector of the coaxial switch.A two-position switch permits you to compare the performance of a test antenna against a second antenna, using the S-meter of your receiver. You must switch back and forth several times before you obtain a reasonable average reading. Fading is a problem when it is rapid and often more than several comparisons are required. When there is a significant difference in the readings you can hear the results especially well if a switch over is made when the signal is at low level. For a strong incoming signal, therefore, the relative performance becomes more obvious when the receiver attenuator is used to set the signal level as recorded on the S-meter at some level S9 or below (with SINPO S 2.5 or below*). The next crucial comparison is to seek out a weak signal almost below readability and compare the two antennas in their ability to give you a barely readable aural output.

Another test that often surprises you is to check out a signal when the radio path is just beginning to open to a distant country. Such a signal often has a crucial low angle of arrival and you can compare the performance of the two antennas under this condition. In fact over a reasonable listening period it is sometimes apparent that first one antenna dominates and then the other.

A three-position switch is often helpful in making such critical comparisons. For example you can then compare quickly three antennas. A typical example would be a quick comparison among a reference antenna, an omnidirectional antenna, and a beam. Or you can compare an antenna against either a horizontal or a vertical reference antenna simultaneously.

(A) Three coaxial lines.

(B) Single-wire fed antennas.

(C) Single-wire fed from receiver.

Fig. 3-1. Methods of connection for quick changeovers among three antennas.

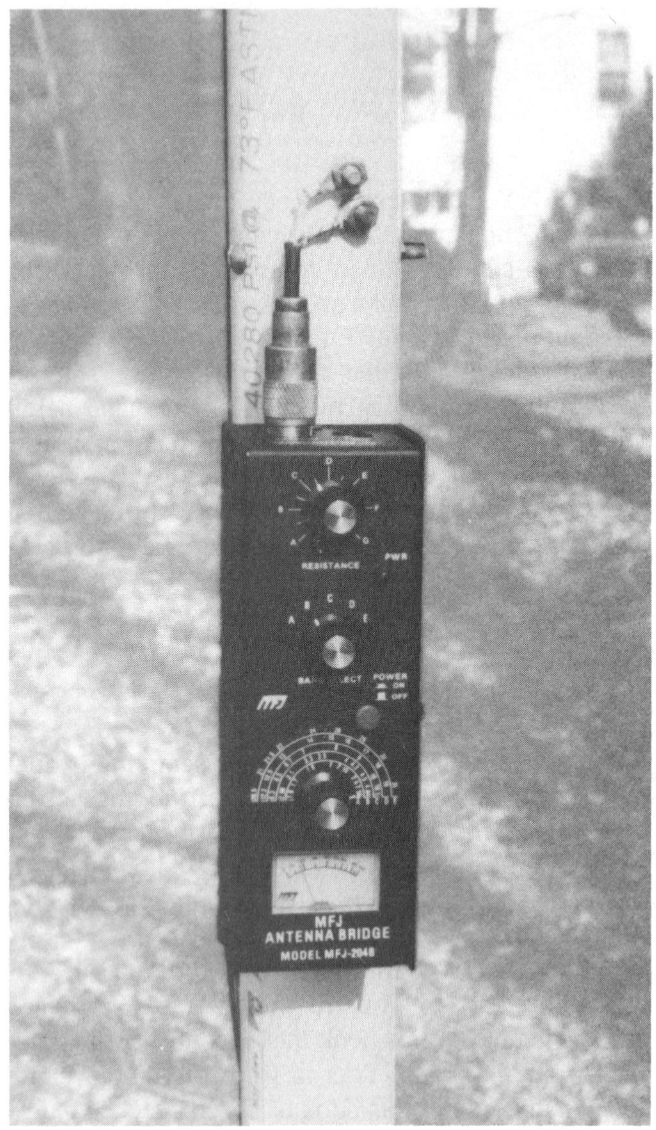

Fig. 3-2. Antenna bridge connected to terminals at base of a 19-meter vertical antenna.

---

*SINPO is a received signal conditions system based on a rating of 0 to 5 for the following characteristics: S—Signal Strength, I—Interference, N—Noise, P—Propagation, O–Overall.

The example in Fig. 3-1A shows how three coaxial lines from three separate antennas can be brought into the radio room to a three-position coaxial switch mounted near the receiver. You can make changeovers as you listen and watch the S-meter. Figure 3-1B shows the same arrangement being used to compare single-wire fed antennas. A short section of line connects the switch output to coaxial-line input of the receiver. The plan of Fig. 3-1C is used when you feed the single-wire input of a receiver.

A battery-powered antenna bridge (Fig. 3-2) is an important adjunct to a test setup because it can be used to check out an antenna right at the antenna terminals. You can measure the resonant frequency and impedance of an antenna at its terminals or with just a very short length of transmission line between the antenna terminals and the meter. The MFJ Enterprises MFJ-204 antenna bridge in Fig. 3-2 consists of four resonant circuits that permit tuning its oscillator frequency between about 1.6 and 35 MHz. The meter reading dips when the oscillator and antenna are tuned to the same frequency. A bridge resistance control is then varied for maximum dip and an associated calibration chart, attached to the rear of the meter, is used to determine the antenna resistance.

Such a bridge is a particular help in trimming an antenna to resonance on a particular band (SWB, ham, marine, etc.) in the HF spectrum. At the same time it measures the antenna feedpoint resistance so you will know how well it matches your transmission line. If the match is acceptable you will obtain identical or nearly the same readings at the receiver end of your line. If you do any amount of antenna trimming you will know it is a joy to be able to carry a small meter right to the antenna and make your necessary adjustments. Make up a simple coaxial adapter like the one in Fig. 3-3 when you must connect meter to bolt/nut antenna terminals.

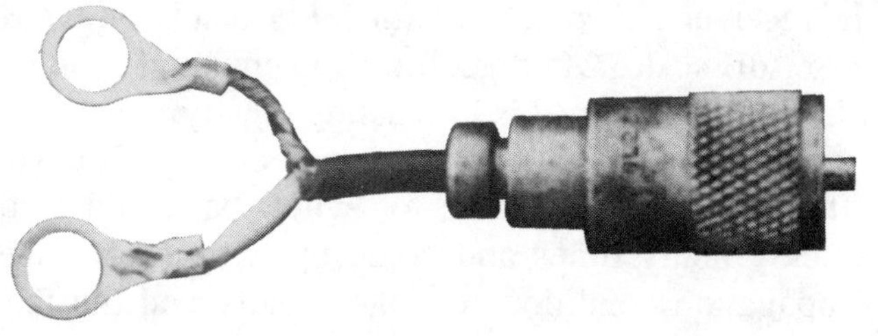

Fig. 3-3. Prepared connector for use with antenna bridge.

3-3. Reference Antennas and Installations

The choice of a good reference antenna and site provides convenience and confidence in checking out a test antenna. Our preferences are the quarter-wave vertical, inverted-V dipole, and loaded vertical. Any one of these types performs well as a HF reference antenna and provides an almost immediate indication of just how well a test antenna is working on a given shortwave broadcast (SWB) band.

Verticals are attractive because they are omnidirectional and provide a reference with just about the same pickup in all directions. Their low-angle sensitivity is good and this is a principal concern for DX reception. However on the lower-frequency bands the conventional vertical is too tall to be practical and it is better to use a dipole type. The inverted dipole is more attractive than the conventional straight dipole because it is easier to change element length when you wish to operate on another band. Also the inverted style is usually more omnidirectional than a regular dipole. Nevertheless it is advisable to position the inverted-V dipole broadside to the signal arrival direction. Since the inverted dipole ends are at low level it is quite easy to reorient the reference when you wish to check it in a different compass direction.

The loaded verticals do well because they reduce the overall height required by a vertical when operating on a low frequency. They are of course omnidirectional like the full-size vertical and need not be changed in making comparisons on signals arriving from other directions. Another advantage of the vertical antenna is that it requires little mounting space and a number can be located in the same general area to form a group of reference antennas. By so doing you allow yourself more space for the erection of various test antennas.

When using an inverted-V dipole as a reference antenna it is possible to use segmented elements and a single installation can be made to operate on two or three adjacent bands. Refer to the end of this chapter for appropriate information.

Knowing true north at the antenna site is important. Knowing the direction of arrival of your DX test signals is also important and is discussed in the next topic. At local noon in any location in the northern hemisphere the sun directs its beam from due south. If you drive a stake into the ground vertically its shadow sets off a line toward the true north at this time. Two corrections must be made. Unless you are on the time zone meridian of your zone your exact noon time is not the same as the clock noon. There is also a time correction based on the fact that the earth's path around the sun is not a perfect circle. Information on these matters is available in various ARRL publications. Also computer services that provide antenna and propagation path data may also make available this information. An appropriate printout would supply you with the clock time representing exact noon at your location for a specified number of days. You need to supply the computer service with your time zone and longitude. Refer to the addresses at the end of Chapter 1.

3-4. Signal Arrival Angles

A beam or any other antenna type that is not truly omnidirectional is best judged by giving some consideration to the bearing angle from true North of the commonly heard DX broadcast signals. Such information can be obtained from an azimuthal map or a printout from a computer service as mentioned in Chapter 1. Prepare an appropriate guide that lists station frequencies, and bearings from your location. These station times and frequencies change over short and long time intervals so you must keep information up to date. You can make any required corrections by referring to current Radio Database International publications, *WRTH Handbook*, or other logging services. Bearings are important in testing an antenna—especially a directional one—and such information should be included in the table.

3-5. Antenna Test Guidance

The first consideration in making an antenna test is to have available a reference antenna for comparison. Such an antenna must be installed properly and perform well on the test band. When it is a resonant type, and most will be such, it should be resonated to the test band with an antenna bridge. You can learn more about the performance of an antenna under test if your reference antenna is doing a good job. Many poor antennas can be made to look good against a poor reference.

The antenna to be tested should be installed and tested with care. An improper installation can make it look bad unfairly as compared with the reference. Use about the same antenna height and the same quality transmission line. Reasonable results can be obtained if the transmission line length differential is held down to something less than 30 feet. On the three higher-frequency shortwave bands keep the differential even less.

When you wish to compare the performance of a test antenna against a reference or against some other antenna, you can do so with a two-position switch. In my own situation I found a good way of comparing two antennas is to compare each against the reference, and then, evaluate how each responded and make my judgment. First erect "antenna one" and check it against the reference. "Antenna one" is now taken down and replaced with "antenna two" in the same general area and connected to the same transmission line. In this manner of comparison you often minimize possible interaction between the two antennas being compared.

If space for isolation is available you can use a three-position switch that will permit a comparison of each in relation to the reference or between each other. This latter plan is sometimes helpful in making listening comparisons among the three relative to the best signal-to-noise or signal-to-interference ratios on a particular station or on a number of signals arriving from the same general direction. Of course you can be jotting down signal level readings at the same time.

Remember that how well you can hear a weak signal and how well a stronger signal delivers a quality reproduction are important in judging an antenna. Sometimes the S-meter reading on a particular signal can be a bit high because the reading is being boosted by interference or QRM. Make your comparison with the receiver RF control set for S9 or below on the S-meter dB scale. Often additional attenuation is an aid when the signal level is fluctuating greatly.

In coming up with a fair S-meter comparison you often find you will have to switch back and forth quickly many times and come up with an average. Propagation conditions change rapidly and the fade cycle on one antenna may not be the same as on the other. Set an average and recheck in five or ten minutes to confirm your results. At other times, when fading is not of the rapid kind, you get the answer almost immediately.

Readings taken when the band is crowded with strong signals near the test signal frequency are not the most reliable. Also strong atmospherics create problems in getting good readings. Quiet daytime readings with signals at moderate to weak levels usually provide the best S-meter comparisons. However listening evaluations are often better and made more conveniently when interference and/or atmospherics are present and the band is crowded. In particular, listening tells more about the performance of a directional antenna than an S-meter reading can disclose. Directional antennas produce a quieter background noise and less interference on the signal arriving from their favored direction(s). Listen and measure if you wish to do a good job in comparing antennas.

3-6. A Test Example

A specific example demonstrates how much you can learn about an antenna with a relatively simple installation. The objective is to compare the performance of a 19-meter sloping one-half wavelength dipole with a one-quarter wavelength vertical used as a reference. If you are installing your quarter-wave vertical at the same time be certain it operates in top form. Construction of such a vertical was covered in Chapter 2. If you have an antenna bridge, use the connection plan of Fig. 3-3. Such a measurement will verify antenna resonance on the 19-meter band. However, if you dimension and install your vertical as detailed in Topic 2-5 you will do fine.

Next install the 19-meter sloper. Details for the sloper construction were given in Topic 2-7. In our test installation the slope was in the direction of Europe. The angle was 45°. A dipole-to-coaxial line connector was used as shown in Fig. 3-4. To connect the antenna to the antenna bridge, a coaxial adapter was needed. This adapter was a double PL-259 to two SO-239 connectors (Radio Shack 278-192). The antenna bridge check was made initially at the top of a short 6-ft (1.83-m) mast to make certain the element lengths were OK before it was erected as a sloper. Check again from the end of a short section of transmission line (Fig. 3-5).

Next an abbreviated table is made ready from the information to be gathered. Once you get into a test routine you can run some extra blank copies of the form that you can use whenever an antenna is to be tested.

Fig. 3-4. Checking dipole resonant frequency and impedance at low height.

Begin with the S-meter comparisons. When you come up with a fair average fill in the blank column for the particular station being received. Averaging suggestions were given in the previous topic. You can't do a complete test immediately because you must schedule readings to match the air times of the various stations. Records of best times for taking readings should be kept up to date. A remarks column for recording listening evaluations may be helpful. In the case of the sloper there was often a definite listening improvement over the vertical antenna in addition to the limited gain in the Europe and Middle East direction even though a station was clear of interfer-

Fig. 3-5. Rechecking the frequency and impedance from the end of a 20-ft. (6.1-m) section of coaxial line after antenna is raised.

ence. It seems that in our location atmospherics and background noises are stronger toward the southwest.

3-7. The 31-, 41-, 49-, and 60-Meter Segmented Dipole

The quarter-wave vertical is an effective reference antenna for 25 meters and the lower wavelength bands. It is more practical to use a dipole reference for the wavelength bands 31 through 120 meters. Additional forms of reference antennas are described in later chapters.

Space saving and band change convenience are the favorable attributes of the inverted dipole. It is also a good performing antenna in its own right. The ends of the inverted dipole are at low height and accessible for making modification or orientation adjustments.

Multiband operation is accomplished by subdividing each leg into four segments with appropriate jumpers as shown in Figs. 3-6 and 3-7. It is a simple matter to change bands by releasing one leg at a time and make the jumper changes needed for operation on a specific band.

Quarter-wave dimensions from the table are 24 ft 1 in (7.35 m), 32 ft 6 in (9.9 m), 39 ft (11.88 m) and

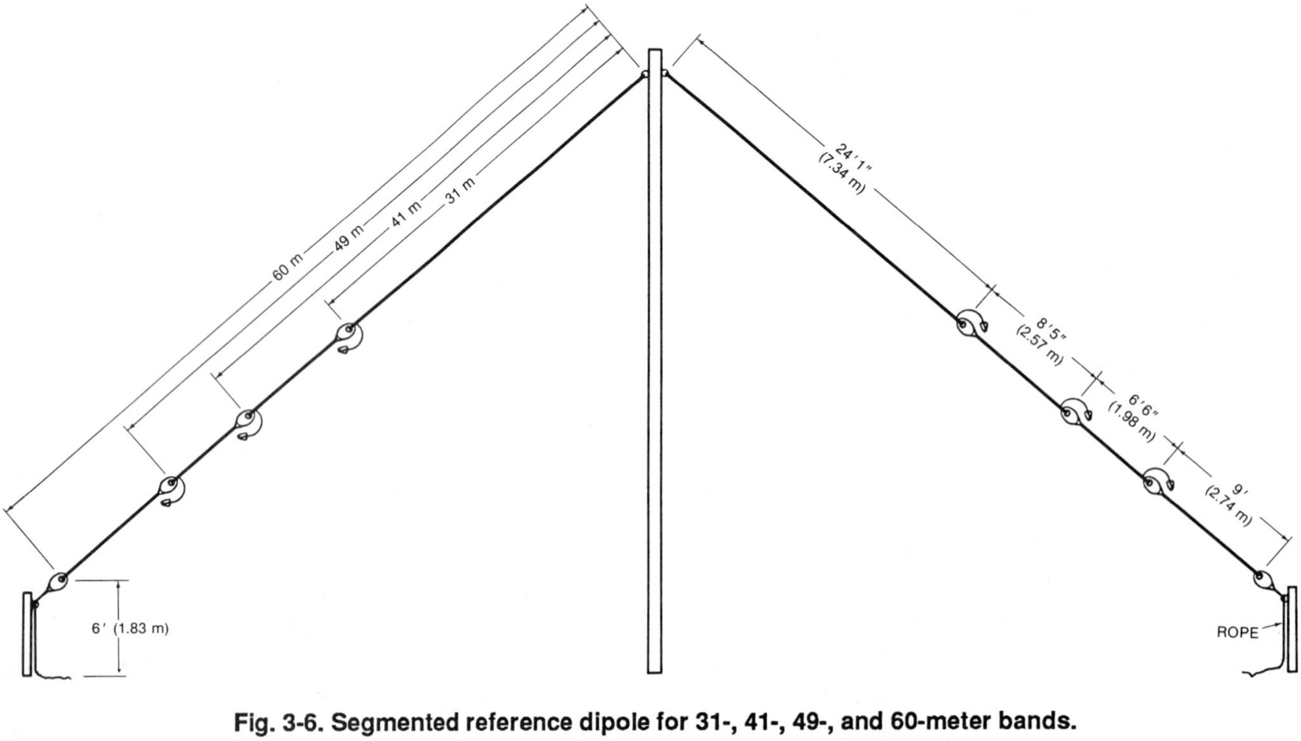

Fig. 3-6. Segmented reference dipole for 31-, 41-, 49-, and 60-meter bands.

48 ft (14.55 m) for the 31-, 41-, 49-, and 60-meter bands, respectively. The segment lengths needed to jumper in the proper sequence for each band beginning at the dipole feed point are 24 ft 1 in (7.34 m), 8 ft 5 in (2.57 m), 6 ft 6 in (1.98 m) and 9 ft (2.74 m). Assemble the antenna. Leave all jumpers open. Now check the performance on the 31-meter band or make a resonance measurement with an antenna bridge. Trim the antenna if necessary. Next drop each leg down and close the jumper nearest the feed point to check out operation on 41 meters. Do the same for 49 and 60 meters. Be certain of your lengths. Listen on each band and you will obtain an idea of how well a properly cut antenna performs as compared to the usual random length antenna.

If you have approximately 120 feet (36.58 m) or so of unobstructed spot on your plot of ground and you enjoy listening on these bands give this antenna a try. Furthermore you can use it for a reference in checking out other antenna types on these bands. If you wish to peak the 75-, 90-, and 120-meter bands you can add additional segments, if space is available.

Fig. 3-7. Wire jumper for segmented dipole.

Chapter 4

Vertical Variations

THE vertical antenna has been neglected and often maligned unfairly as a poor shortwave receiving antenna. Actually its performance is good and often exceptional. Furthermore little mounting space is needed and the vertical can be an attractive alternative when lot size is a problem. Verticals can be loaded and made to perform well despite lengths that are much shorter than a quarter wavelength. Such arrangements are even effective on the tropical and medium-wave bands.

4-1. Loaded Vertical Antennas

There are four common ways of loading a vertical (Fig. 4-1). In Fig. 4-1B a base loading coil is used. This coil adds the inductance needed to bring the capacitance of a short antenna into resonance, presenting a low impedance at the base feed point. Center loading (Fig. 4-1C) produces the same results, often with a smaller inductance. A helical winding (Fig. 4-1D) can also bring a short antenna into resonance. This style can be made to operate with good sensitivity over a somewhat wider bandwidth. A zigzag or laced vertical antenna (Fig. 4-1E) affords an easy and effective technique for shortening physical length when using a nonconducting mast, such as PVC pipe.

The various loading procedures can be combined in two or more combinations to obtain a number of special arrangements for shortwave broadcast listening and ham operations. All of the loading arrangements must be adjusted to peak their performance. An antenna bridge or at least a reference antenna is a definite aid in adjusting such a loaded vertical. Acceptable results can be obtained by duplicating the construction information in the topics that follow although the use of an antenna bridge will help you obtain the very best operating conditions.

4-2. Base-Loaded 31-Meter Vertical

The length of a quarter-wave vertical for operation on the 31-meter band from the dimension tables is 24 ft 1 in (7.35 m). An example of how a base loading coil can shorten the overall length of a 31-meter vertical is shown in Figs. 4-2 and 4-3. The vertical and its associated coil occupy a 15-ft 3-in (4.65-m) length on the 2-in (5.08-cm) ID PVC piping mast. Three 24-ft 9-in (7.54-m) radials are used as shown in Fig. 4-4. The three radials can be dropped down the mast and buried just under the surface. If you are pinched for space, radials can be folded back and forth as shown by the dashed curve in Fig. 4-4.

The base-loaded vertical was mounted on two telescoped sections of PVC piping. Refer to Chapters 2 and 3 for antenna construction and test details. The vertical wire is 15-ft (4.57-m) long and the coil consists of eight turns of vinyl covered No. 16 wire close

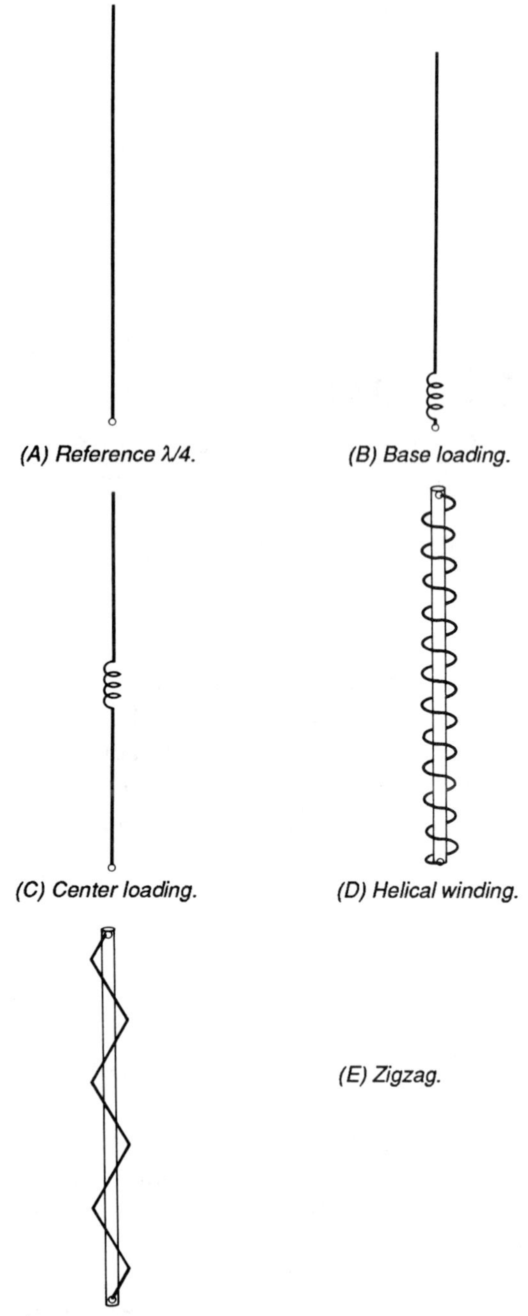

Fig. 4-1. Vertical antenna loading methods.

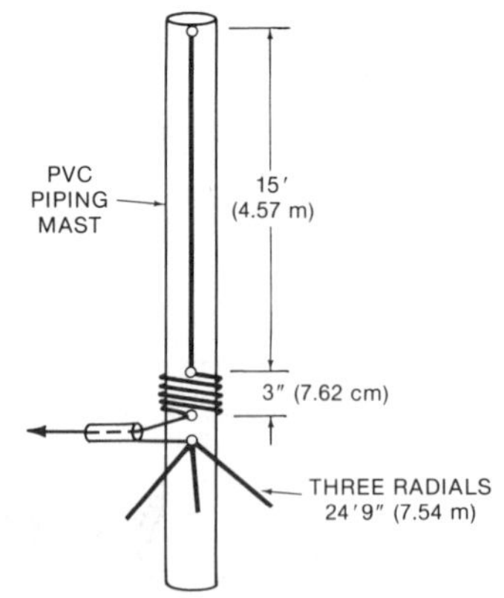

Fig. 4-2. The 31-meter base-loaded vertical.

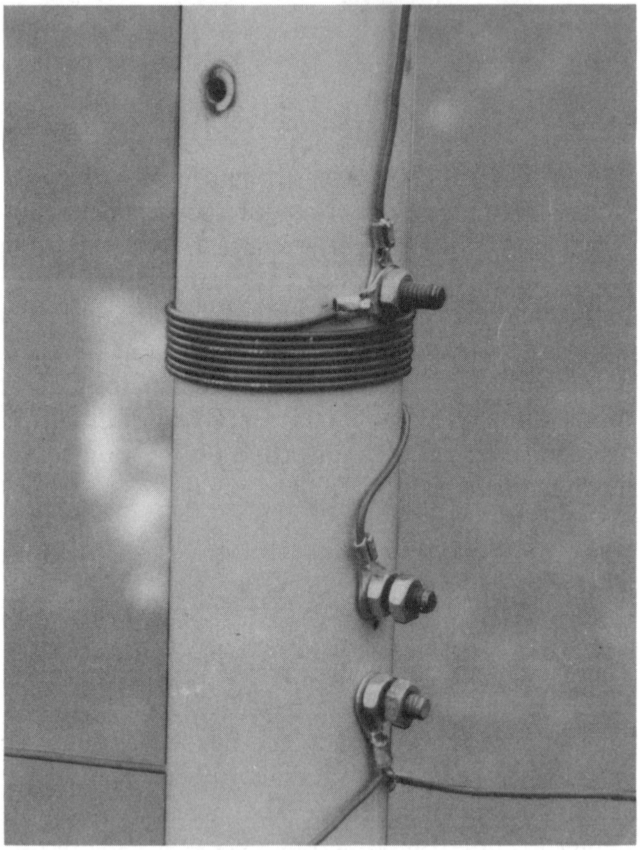

Fig. 4-3. Coil of the base-loaded vertical.

wound. Details of the construction can be seen in Fig. 4-3. The coaxial line connects between the bottom of the coil and the junction of the three radial wires. The antenna resonated at 9.5 MHz with excellent results on 31 meters and good pickup on both 41 and 25 meters as well.

If you include an arrangement for connecting a jumper across the coil, the antenna also operates well on 19 meters and has good pickup on 16 and 21 meters. The jumper point is accessible from ground level and there is no connection problem. You can anticipate good results on bands 16 through 41 meters with this very simple base-loaded antenna that requires so little space. A second set of three radials is required for higher band operation. These radials are 15 ft 8 in (4.78 m) in length and should be positioned half way between the 31 meter radials.

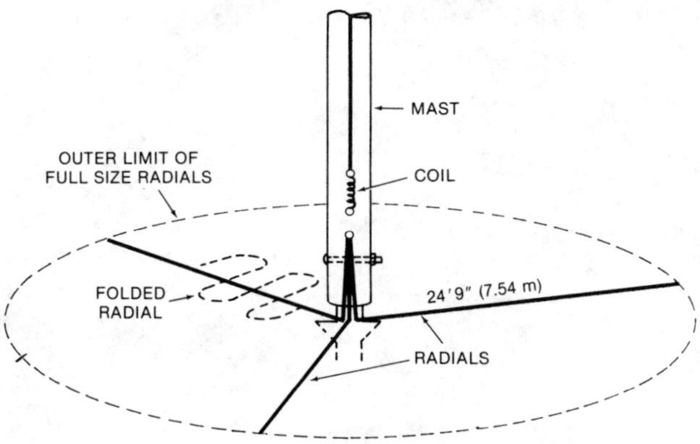

Fig. 4-4. The radial system.

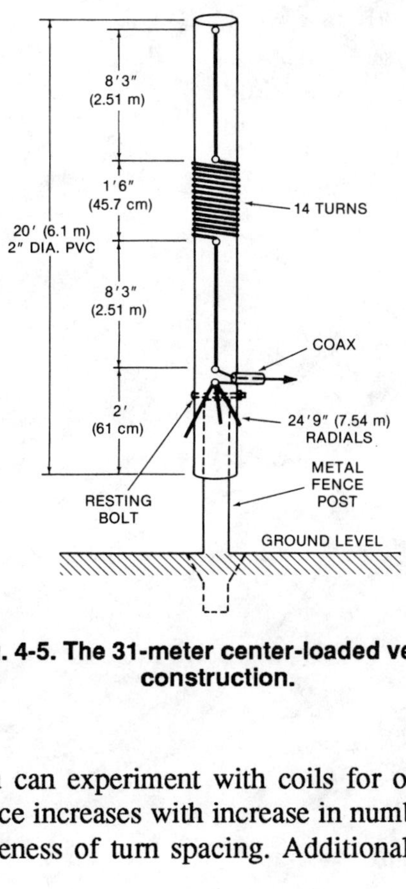

Fig. 4-5. The 31-meter center-loaded vertical construction.

You can experiment with coils for other bands. Inductance increases with increase in number of turns and closeness of turn spacing. Additional coil plans follow.

4-3. The 31-Meter Center-Loaded Vertical

A space-saving vertical for the 31-meter band can be mounted on a twenty-foot length of 2-inch ID PVC piping as shown in Figs. 4-5 and 4-6. The overall length of the vertical antenna wire (No. 16 vinyl covered) is about 28 ft (8.53 m). It is subdivided in two 8-ft 3-in (2.51-m) lengths plus the piece used to wind the center-loading coil which extends over a centered length of 1 ft 6 in (45.7 cm). After the mast was erected it had to be lifted twice off the metal fence post to determine the exact number of turns (14) which would center the resonant frequency on the 31-meter band. The ground system consisted of three 24-ft 9-in (7.54-m) radials. As mentioned before these radials can be laid on the ground or buried one or two inches beneath the surface. Again, if you must mount in a confined space the radials can be laid in a serpentine manner as shown by the dashed curve of Fig. 4-4. The ends of the radials are soldered to a solder ring which connects over the ground terminal as shown previously in Fig. 4-3. All of this must be constructed before you can really check out the resonant frequency.

Fig. 4-6. Center-loading coil wrapped around PVC pipe.

The dimensions of Fig. 4-5 provide good performance over the entire 31-meter band. The antenna also performs well over the 25-meter band and provides acceptable results on some of the other SWB bands too. Coils can be wound for other bands. Refer to the topics that follow.

4-4. The 31-Meter Helical-Wound Antenna

A helical winding permits an alternative arrangement for shortening the height of a quarter-wave vertical and, at the same time, provides good performance. The example of Figs. 4-7 and 4-8 demonstrate how a 31-meter helix confines the normal 24-ft 1-in (7.35-m) quarter-wave length of a vertical on this band to 13-ft (3.96-m) height. Vinyl-covered No. 16 wire is wound around the PVC mast with a spacing of approximately 1.75 in (4.45 cm).

An approximate rule-of-thumb for winding a helical antenna is to start with a length of antenna wire that is about twice the length of a quarter wave on the operating frequency. In the example a length of about 50 ft (15.24 m) was selected. The final version was cut a bit shorter to resonate at 9.7 MHz as measured on the antenna bridge.

A helical winding can be used with other forms of loading to construct other types of short and versatile antennas that require little mounting space.

4-5. Zigzag Vertical

A fourth method of shortening the physical height of a vertical is to zigzag or lace the antenna wire from side to side as it proceeds down the mast as shown in Figs. 4-9A and 4-10. Adjacent eyebolts or simple screw-eyes are mounted on opposite sides and spaced at 1-ft (30.5-cm) intervals down the mast. In the example shown a 16-ft 3-in (4.95-m) length of wire resonated the vertical on the 21-meter band. The total antenna length was approximately 13 ft 4 in (4.06 m). A resonant quarter wave on the 21-meter band is 17 ft (5.18 m). Hence the zigzag technique shortened the overall antenna length by 3 ft 8 in (1.1 m).

A double zigzag scheme can be used to resonate the system at even a lower frequency using the same

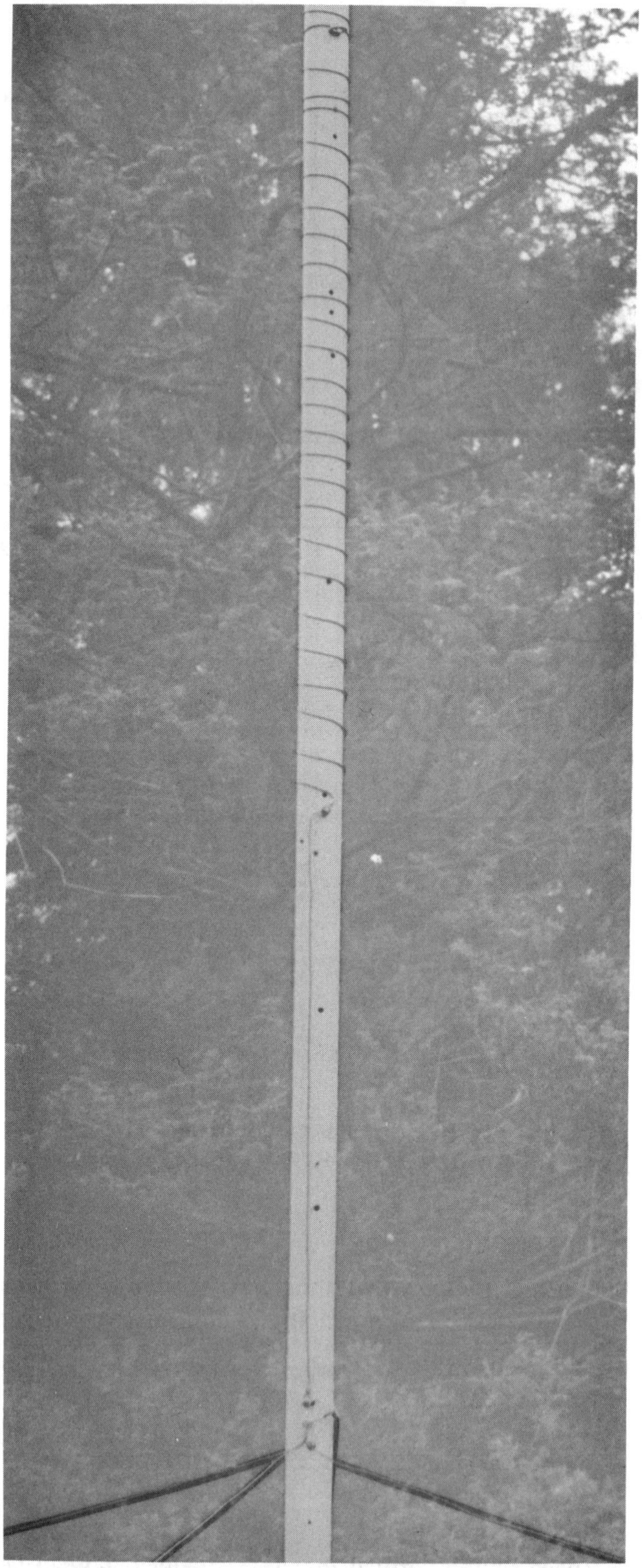

Fig. 4-7. The 31-meter vertical showing helical winding.

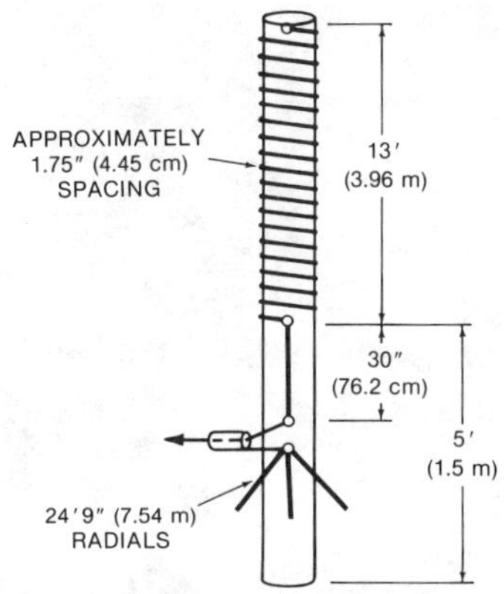

Fig. 4-8. The 31-meter helical winding construction detail.

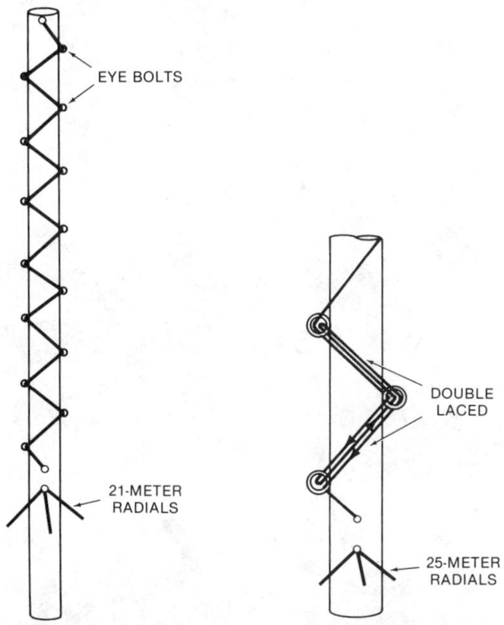

(A) 21-meter band. *(B) 25-meter band.*

Fig. 4-9. Zigzag loaded verticals.

overall length. In Figs. 4-9B and 4-11 the antenna wire is doubled back two positions before dropping down the same path again to the antenna terminal. By so doing the resonant frequency is dropped to 25 meters. A quarter wavelength straight vertical on this wavelength is 19 ft 10 in (6.04 m).

4-6. Base-Loaded Multiband Vertical

A small coil with taps mounted at the base of the 19-meter vertical antenna described in Topic 2-5 can peak the performance of the vertical on 21 and 25

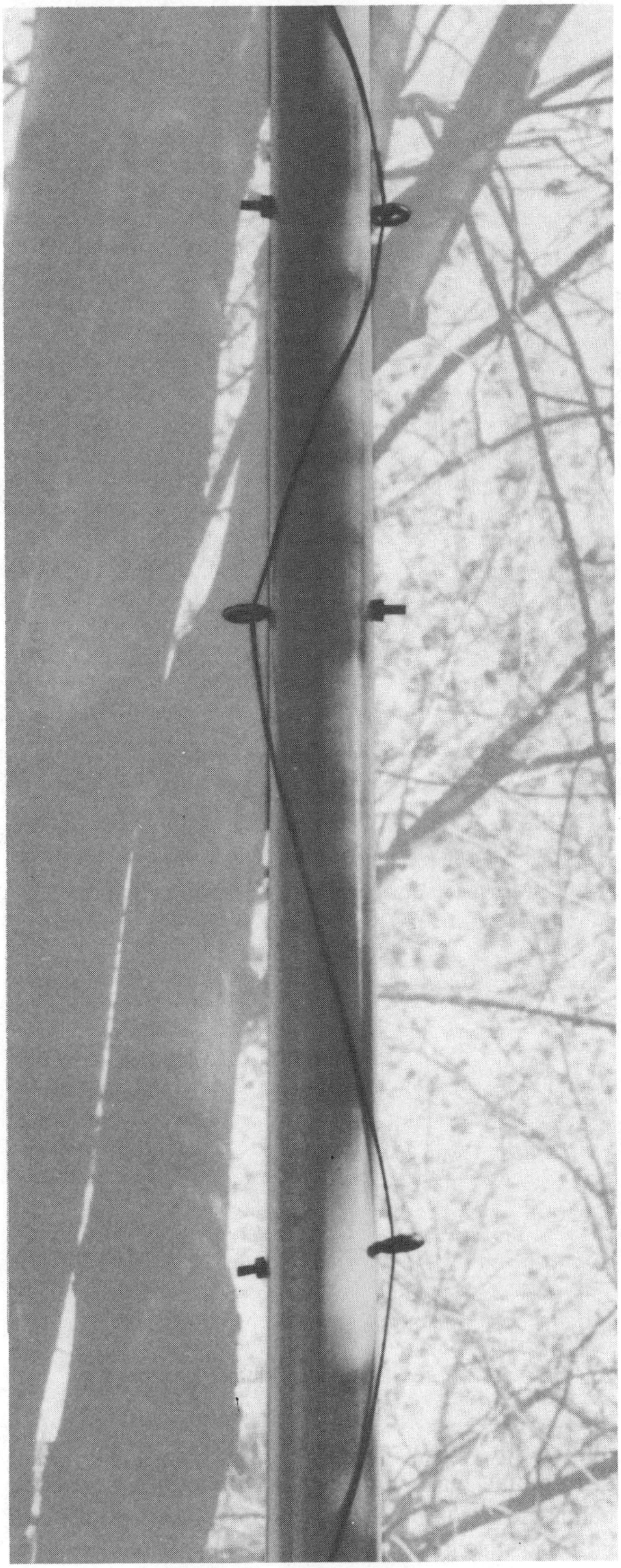

Fig. 4-10. Zigzag wire and eyebolt to guides.

Fig. 4-11. Double back-and-down lacing to lower resonant frequency.

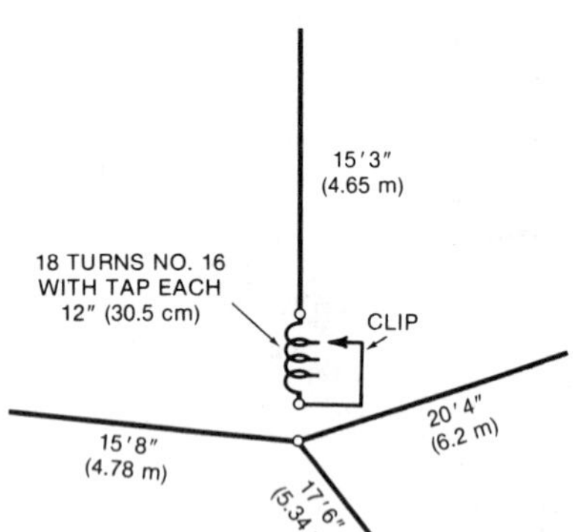

Fig. 4-12. Base-loaded multiband vertical plan.

meters. In this arrangement (Figs. 4-12 and 4-13) three radials, each a different length to match the three bands are used. By adding additional turns to the coil you can also peak performance on the 31- and 41-meter bands. The antenna peaked well on the latter

Fig. 4-13. Mast and base loading coil.

two bands without longer radials. You may wish to try longer radials despite the added space required.

The coil (Fig. 4-14) consists of 18 close-wound turns with taps positioned every 12 in (30.5 cm). The correct tap for each band can be found by using a receiver, two-position coaxial switch, and reference

Fig. 4-14. Detail of the tapped loading coil.

antenna or, more simply, with the antenna bridge. This compact base-loaded antenna performance will surprise you. Such tapped-coil verticals do well as reference antennas.

4-7. Universal Loading Coil and Matcher

A tapped base-loading coil and two clip jumpers provide a versatile combination that can be used with a variety of vertical and single-wire antenna types. Such an assembly can be used to improve the low frequency performance of a short antenna, to obtain better results indoors and in other restricted space locations, to serve as a part of a good-performing reference antenna, and to save considerable time and cost in the

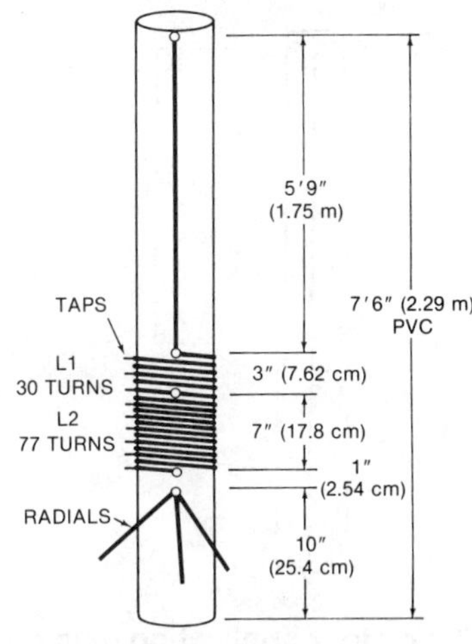

Fig. 4-15. Loading coil assembly data.

Fig. 4-16. Construction details of the two loading coils.

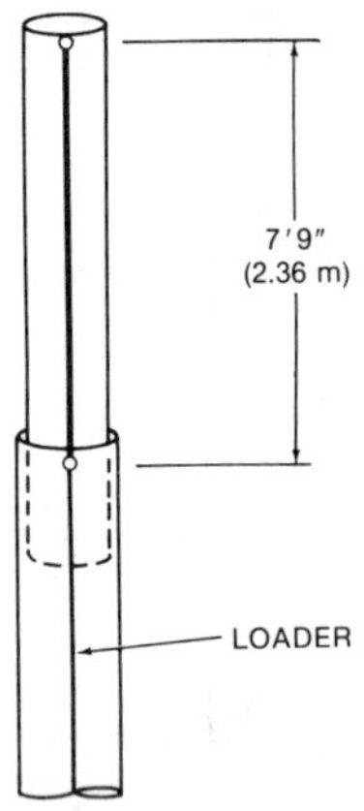

Fig. 4-17. Outdoor application of loader with a 19-meter vertical.

planning and testing of antennas for a variety of services. Its applications include shortwave and medium-wave listening as well as ham operations. The assembly can be used outdoors, indoors, and in portable situations. Antennas built around this unit are covered in later chapters.

The assembly is mounted on a 7-ft 6-in (2.29-m) length of 2-inch (5.08 cm) ID PVC piping as shown in Fig. 4-15. This length was chosen because it can be used indoors and it is transportable in a station wagon. There are two separate windings (L1 and L2) close wound with No. 14 vinyl-covered wire. Figure 4-16 shows the details. Coil L1 is made from a 20-ft (6.1-m) length of wire with taps spaced 1-ft (0.305-m) apart; coil L2 a 50-ft (15.24-m) length with taps separated by 10 ft (3.05 m). A short piece of wire with clips is attached to the bottom of each coil with a length just sufficient to reach the furthermost tap and also jump the individual coil. Coil L1 has 30 turns and coil L2 has 77 turns.

In an outdoor test of the loader an additional 7-ft 9-in (2.36-m) length of antenna wire was added to the top of the 5-ft 9-in (1.75-m) assembly wire (Fig. 4-17). You can do so by telescoping and bolting a 1.5-in (3.8-cm) ID pipe into the top and, making wire connection at the top as shown. The added wire length resonated the antenna on 19-meters when both coils L1 and L2 were jumped. Three 19-meter radials were connected. It was possible to resonate this simple vertical on all SWB bands from 19 through 120 meters.

When coil L2 was jumped, the taps on L1 found resonances on the 21- through 60-meter bands. When coil L2 was jumped to its tap 2 positions from the top, the taps on L1 were used to find resonances on 75 and 90 meters. When coil L2 was opened completely for operation (no short), a proper tap on L1 found resonance on 120 meters. The antenna bridge described in Chapter 3 was of great assistance.

The bandwidth of the loader and optimum tap positions vary with the characteristics of the antenna and ground system with which it is associated. More details are given later in this book as it is applied in other situations and with other antenna types.

Chapter 5

Dipole and Longwire Variations

IS the dipole strictly a one-band antenna? Not always when there are suitable harmonic relations. Also, dipoles can be arranged in two or more combines to obtain good multiband performance. These and other variations are covered in this chapter.

5-1. Introduction

First consider harmonic relations. When each leg of a dipole is one-quarter wavelength long, the voltage and current distribution is as shown in Fig. 5-1A. At each end there is maximum voltage and minimum current (high impedance). Exactly one-quarter wavelength from each end there is a low impedance as the voltage becomes minimum and the current maximum. A low-impedance transmission line can be attached here (at the center) and there is efficient transfer of signal current to the line.

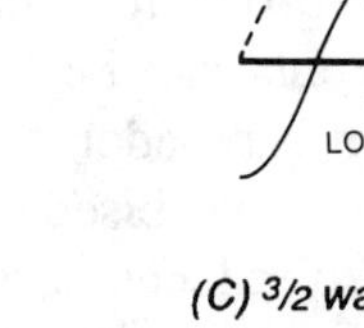

(A) Half wave dipole.

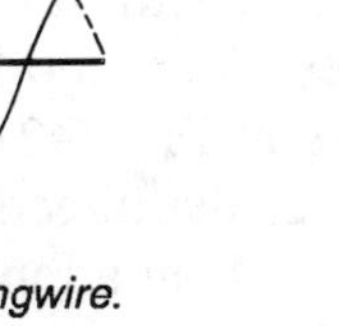

(B) Full wave.

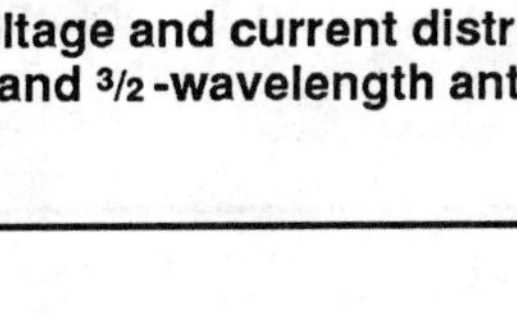

(C) 3/2 wavelength longwire.

Fig. 5-1. Voltage and current distribution for 1/2-, 1-, and 3/2-wavelength antennas.

Notice in Fig. 5-1B what happens to the voltage and current distribution when an additional one-quarter wavelength is added to each dipole segment. Each leg is now a half wave in length and the center location is at a high impedance point which is an unfavorable situation for transferring signal to a low-impedance line. Observe in Fig. 5-1C when an additional quarter wavelength is added to each segment the center returns to a low-impedance point once again and is suitable for the transfer of signal to a low-impedance line. Such an antenna is 3/2 wavelength long and is made of two 3/4-wavelength segments. A relatively low impedance exists at the center feed point whenever each dipole leg is an odd multiple of a quarter wavelength long.

Examples of such antennas are given in Fig. 5-2. The actual lengths given are based on 19-meter operation. Refer also to the 3/4-wavelength column of the shortwave broadcast band dimension table of Chapter 11. Here you can find the length required to set up a 3/2-wavelength antenna on any one of the bands. This column is repeated in Table 5-1. Table 5-1 also gives length values that correspond to leg lengths of 5/4 and 7/4 wavelength on the SWB bands. For each entry the

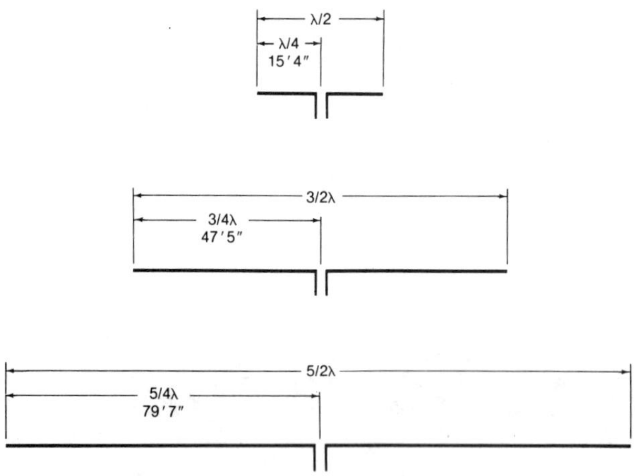

Fig. 5-2. Low impedance center-fed antennas for 19 meters.

Table 5-1. Longwire Antenna Lengths

Freq. (MHz)	Meters Band	3/4 λ	5/4 λ	7/4 λ
2.4	120	302' 92.2	508' 154.7	713' 217.2
3.3	90	220' 67.06	369' 112.5	518' 158
3.95	75	184' 56.03	308' 94	433' 132
4.9	60	148' 45.16	249' 75.76	349' 106.4
6	49	121' 36.88	203' 61.87	285' 86.87
7.2	41	101' 30.74	169'2" 51.56	237'6" 72.39
9.7	31	75' 22.81	125'7" 38.27	176'4" 53.73
11.8	25	61'6" 18.75	103'2" 31.46	144'11" 44.17
13.7	21	53' 16.15	88'11" 27.09	124'10" 38.04
15.3	19	47'5" 14.46	79'7" 24.26	111'9" 34.06
17.7	16	41' 12.5	68'10" 20.97	96'7" 29.45
21.6	13	33'7" 10.25	56'5" 17.19	79'2" 24.13
25.8	11	28'2" 8.578	47'2" 14.39	66'3" 20.20

$3/4\ \lambda_{Feet} = 726/f_{MHz}$
$3/4\ \lambda_{Meters} = 221.3/f_{MHz}$

$5/4\ \lambda_{Feet} = 1218/f_{MHz}$
$5/4\ \lambda_{Meters} = 371.2/f_{MHz}$

$7/6\ \lambda_{Feet} = 1710/f_{MHz}$
$7/4\ \lambda_{Meters} = 521.2/f_{MHz}$

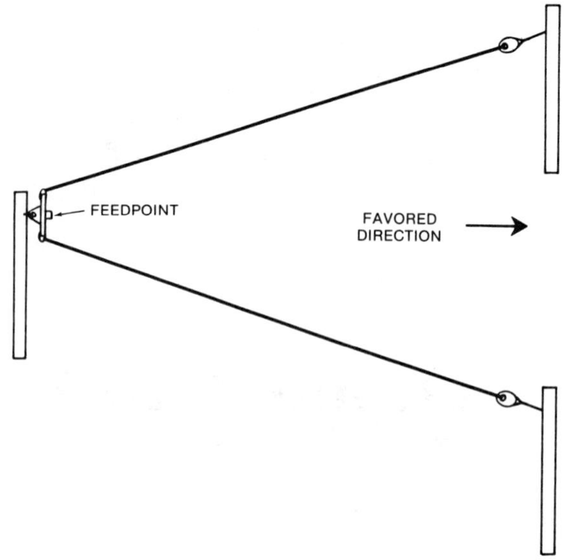

(A) Long segments tilted forwards.

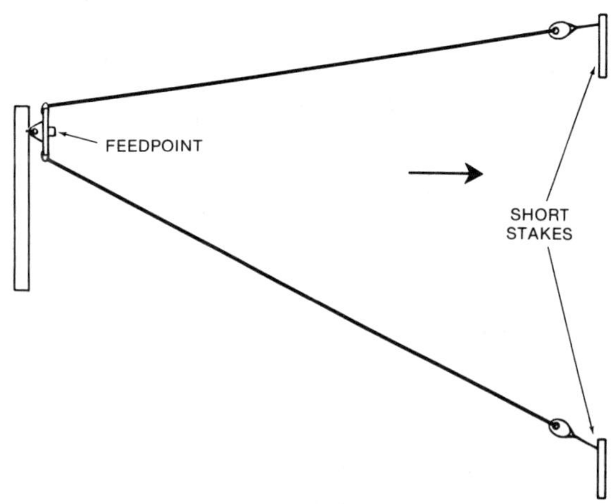

(B) Tilting wires toward ground.

Fig. 5-3. Gain and bandwidth improvement methods for longwires.

top row is the length in feet and the bottom row the length in meters.

Such antennas provide gain off the leg ends and less pickup broadside to the antenna wire. The longer the antenna wire is in terms of the number of odd quarter wavelength the sharper and higher becomes the end sensitivity while the broadside sensitivity declines. Also, the long segments can be tilted forward as in Fig. 5-3A to obtain a broader directional pattern in the direction of a line that bisects the two wires. Such an arrangement also reduces back pickup and concentrates more of the sensitivity forward. Additional bandwidth can be obtained by also tilting the two wires toward the ground as in Fig. 5-3B. A number of these antennas are discussed in this chapter. Let's begin with a simple 3/2-wavelength sloper.

5-2. The 3/2-Wavelength Sloper

A 3/2 wavelength 19-meter sloper of Fig. 5-4 was constructed and oriented towards Europe. The antenna started out with a 47-ft 5-in (14.5-m) leg length suggested in the table. However to center the performance in the 19-meter band it was necessary to cut 6 inches off of each leg. Learn to live with the influence of variables. The height of the mast was about 22 ft (6.7 m).

The sloper directivity was quite apparent. Although there were a couple of inconsistencies that were not unexpected because of the multiple lobe pattern of a longwire antenna. Also the vertical angle of arrival can change with propagation conditions. Therefore the relative vertical patterns of a 3/2-wavelength sloper and of a one-quarter wavelength vertical reference will influence the delivered signal levels. In general, the 3/2-wave sloper sensitivity and directivity are somewhat better than those of a half-wave sloper. Side and back pickup was definitely poorer, indicating the sloper would do well in cutting back on the effects of interference from stations to the side and rear. This was verified very quickly on 15300 kHz when a signal from France became at least readable when Cuba was on the frequency at the same time. On the quarter-wave reference antenna the Cuban station was strongly dominant. On 15375 kHz, Spain and WYFR (USA) interacted but Spain was much stronger than WYFR on the 3/2-wavelength sloper. On 15140 kHz, Chile dominated on the vertical; USSR, on the sloper. The three examples demonstrate how a little extra gain in the reception of a desired station and less gain of an interfering station to the rear provide the benefit of a directional antenna. Remember that the low end of a sloper is accessible from the ground and the antenna can be easily swung around to favor another direction when you wish to do so.

Although the antenna was designed for 19-meter operation as a 3/2-wavelength sloper it is capable of fine performance on 60 meters where it operates as a half-wavelength sloper. Note from the SWB dimension chart in Chapter 11 that a quarter wavelength on

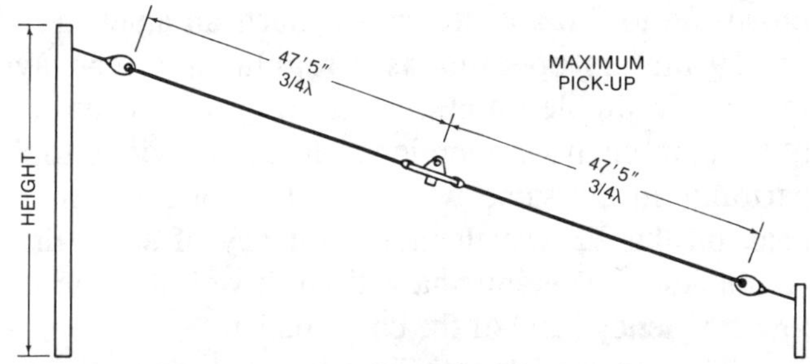

Fig. 5-4. A 3/2 sloper for 19 meters.

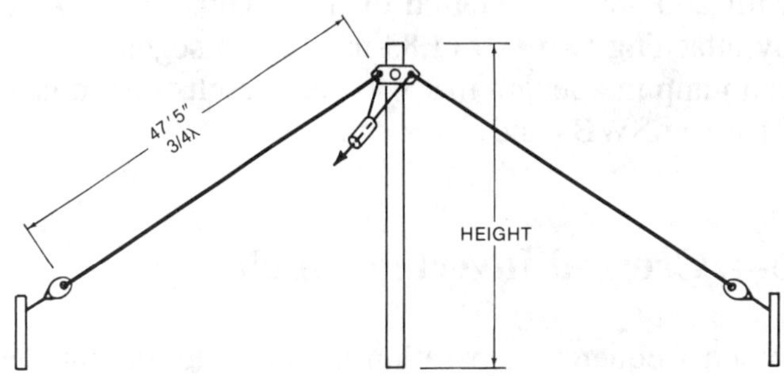

Fig. 5-5. A 3/2 wavelength inverted longwire antenna.

60 meters and 3/4 wavelength on 19 meters are about the same physical length. The frequency of the 19-meter band is almost the third harmonic of the 60-meter band frequency. Thus the antenna performs very well on both frequencies because the impedance at the center of the antenna is low, ready to transfer signals to a low-impedance transfer line.

There are five other third harmonic relations to consider on the SWB frequency assignments. For example a 3/4-wavelength segment on 25 meters will operate as a one-quarter wavelength leg on 75 meters, and a 3/4-wavelength leg on 31 meters operates as a one-quarter wavelength leg on 90 meters. Other relationships are 41–120 meters, 16–49 meters, and 13–41 meters. These relationships are useful in planning effective multiband antennas that perform very well on more than just two bands.

5-3. The 3/2-Inverted-V Antenna

A less directional antenna can be installed with the inverted-V 3/2-wavelength configuration of Fig. 5-5.

There is directivity off both leg ends and less response broadside to the antenna wires. Such an antenna cut for 19 meters operates as a regular quarter-wave inverted-V dipole on 60 meters. Any one of the five other combinations mentioned in the previous topic performs in the same way. Trim the antenna wire to peak on the 3/2-wavelength frequency of any paired combination. The antenna will work well then on the low-frequency band of the combination.

The 3/2 wavelength inverted-V does well as a bidirectional antenna. In my location such a bidirectional pattern provides fine Australia/Spain reception with 260°–80° orientation of its antenna wires. Also by attaching two 6-ft (1.83-m) add on segments you can jump in a length that will peak results on the new 21-meter SWB band.

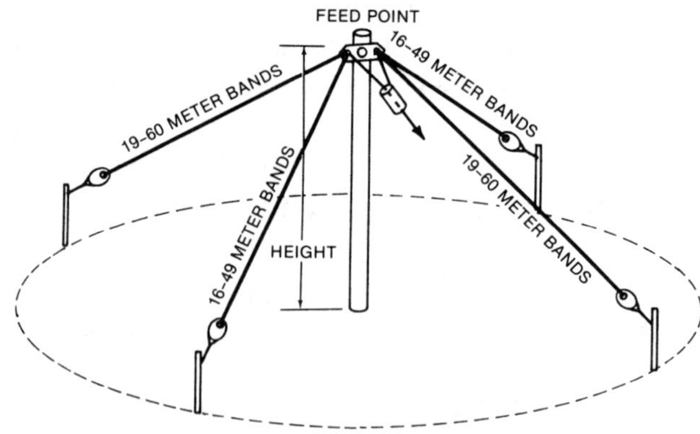

Fig. 5-6. Four-band crossed inverted dipoles.

5-4. Crossed Inverted Dipoles

When frequency separation is adequate among the frequencies to be peaked by an antenna, two dipoles can be interconnected and crossed and fed to the same transmission line with minimum interaction. Thus, both dipoles operate with good efficiency. Frequency separation among the SWB assignments usually permit this mode of operation.

Good results were obtained from the crossed dipoles arrangement of Fig. 5-6. Its dimensions are based on the 60–19 and 49–16 pairing. After installation do any end trimming, if needed, to peak the results on 16 and 19 meters. In this arrangement you have a four-band antenna on 16, 19, 49, and 60 meters. Acceptable pickup is also obtained on the 25-, 31-, and 41-meter bands.

5-5. Quarter-Wave Sloper

The quarter-wave sloper is an effective antenna when space is not available for the erection of a half-wave sloper. Space can be a problem for operation on the tropical broadcast bands and even on the 41- and 49-meter bands if your lot size is very small. This short sloper has but limited directivity but its performance is fine considering its short length. Typical lengths for 41, 49, and 60 meters are given in Fig. 5-7A. As you can determine from the dimension tables of Chapter 11 the wire length for a 120-meter version is only 98 ft (29.71 m).

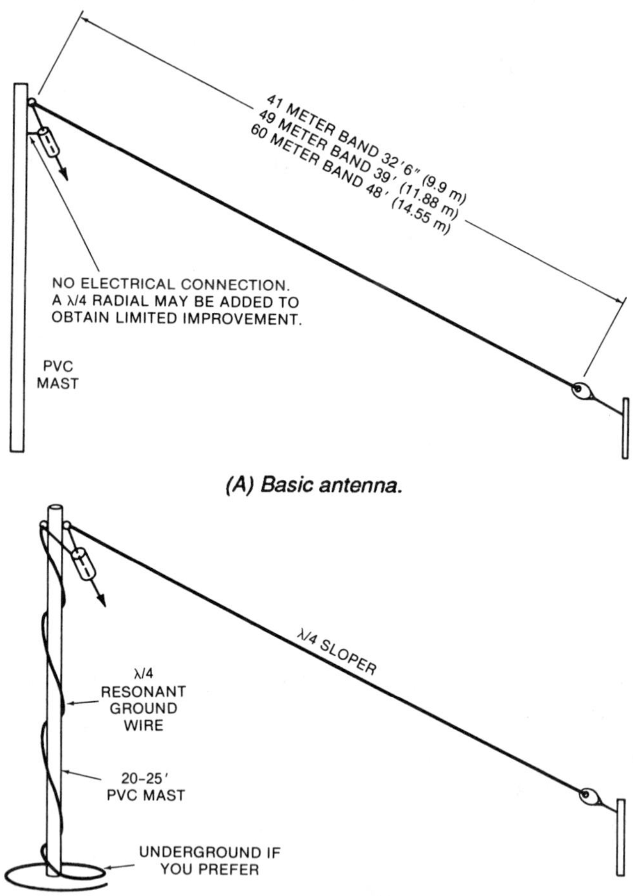

(A) Basic antenna.

(B) Addition of resonant ground wire.

Fig. 5-7. The basic λ/4 sloper and ground-wire addition.

The construction is very simple. The inner conductor of the coaxial line connects to the high end of

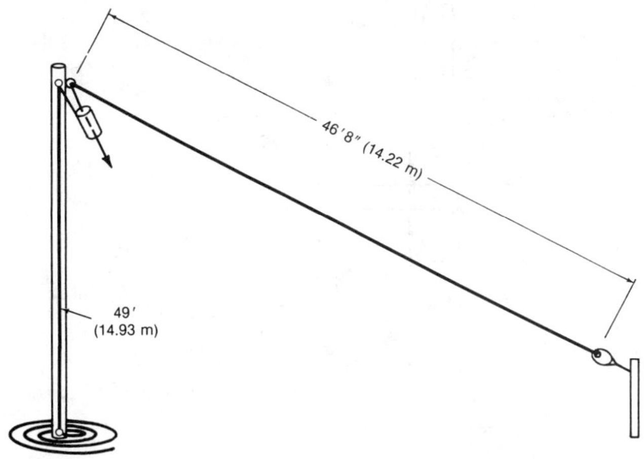

Fig. 5-8. Good compromise antenna for 41, 49, and 60 meters.

the sloper wire. In the basic style of this antenna the coaxial braid is not connected to any antenna wire but can be connected to a terminal mounted at the top of the mast for support purpose only.

The operation of the quarter-wave sloper can be peaked by attaching the braid to a resonant ($\lambda/4$) ground wire that is taped down the mast as shown in Fig. 5-7B, and then looped around the base of the mast in a spiral. You can bury the wire an inch or two beneath the ground level. One of the problems of feeding a quarter-wavelength segment of wire is that the performance of the installation is influenced by the overall length of the coaxial transmission line that connects it to the receiver. The use of the resonant ground wire or a simpler ground system reduces this transmission-line length problem.

The arrangement of Fig. 5-8 is a good performing combination for the 11-, 49-, and 60-meter bands. The antenna wire was cut a bit shorter than the specified 60-meter quarter wavelength. In combination with the 60-meter resonant ground wire, the antenna does very well on all three bands considering the limited space that is required.

5-6. Dipole and 300-Ohm TV Line

Flat 300-ohm TV line is a good choice for conveying HF signals to a SWB receiver. Even good quality TV "ribbon" line is inexpensive as compared to coaxial line and requires only low cost connectors. Does it

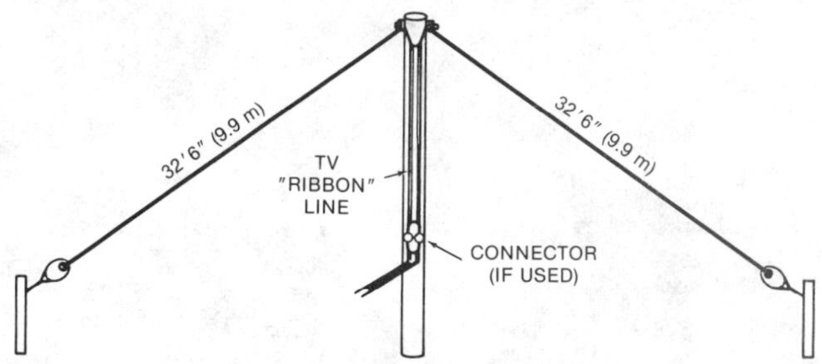

Fig. 5-9. Inverted dipole with 300-ohm TV feedline.

pick up more noise than coaxial line? On rare occasion this may be objectionable. However in most practical situations it is the antenna that is picking up the noise and not the line. What about the mismatch? Some mismatch does occur at antenna resonance but signal amplitude usually remains high because there is less line attenuation as compared to that of a coaxial line. Off resonance the situation improves. The antenna impedance rises off resonance and the line match is better than obtainable with the coaxial line. Stated another way the antenna installation has an improved bandwidth. Some additional improvement results when your receiver includes a high impedance antenna input of several hundreds of ohms. Are you willing to give it a try? When you do so always use a reference antenna for comparison.

A good installation is shown in Fig. 5-9. The dipole has been cut for the 41-meter band and you can anticipate dipole-like performance on the 31-, 41-, and 49-meter bands. Good results can also be expected on some of the other bands too, especially 19 and 13 meters.

In the construction example two dipole bolt/nut terminals are mounted at the very top of a 20-ft (6.1-m) PVC mast as shown in Fig. 5-10. Eye-ring solder lugs are attached to the antenna wire ends and to the two ends of the 300-ohm TV line. A 10-ft (3.05-m) section of line is taped down the mast from the terminal to chest level. Here a TV-line connector is attached to the ends. A longer section of line is then used to convey the signals back to the radio room.

5-7. A "Flexo" Switch Antenna

A "Flexo" switch adds versatility to many antenna styles that can increase signal level and, on occasion,

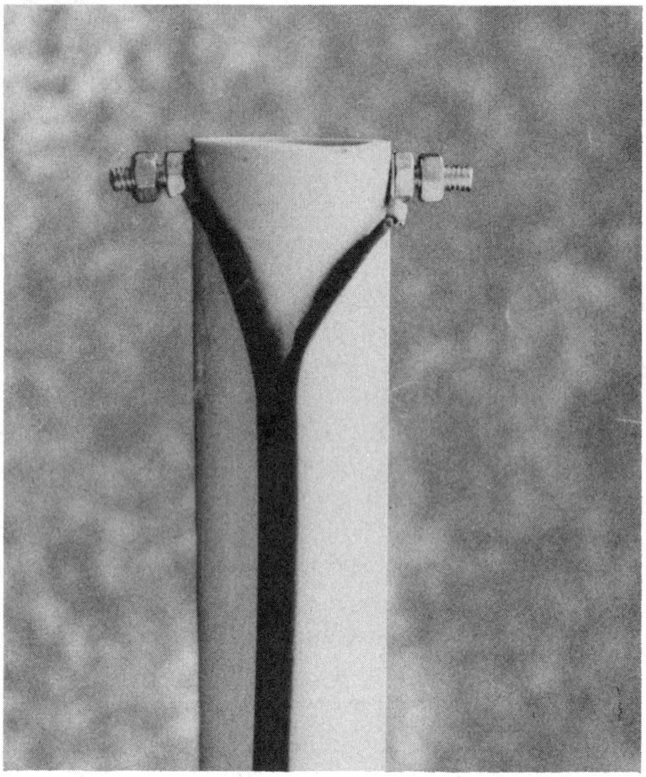

Fig. 5-10. Antenna terminals at top of mast showing TV-line connections.

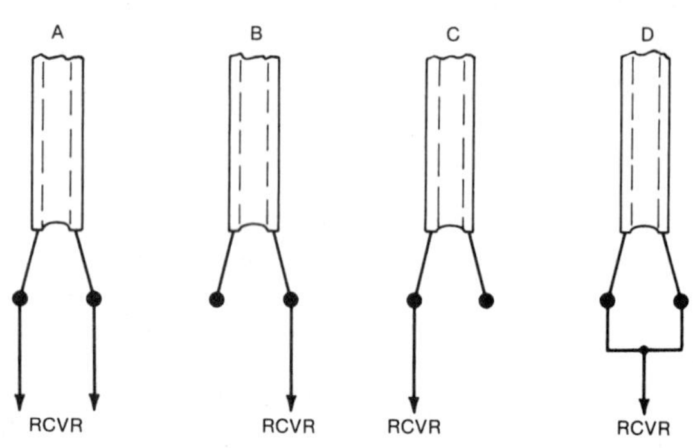

Fig. 5-11. Four ways of connecting TV line to receiver.

provide some reduction of on-channel and adjacent-channel interference as well as noise. The flexo can also improve performance on bands for which the antenna has not been specifically cut. An example is the antenna of Fig. 5-9. A flexo switch will extend its performance to 15 meters as well as the 75-, 90-, and 120-meter tropical bands. Also there is some improvement on the bands that extend between 11 and 21 meters. In fact the installation does quite well over the spectrum between 2 and 30 MHz.

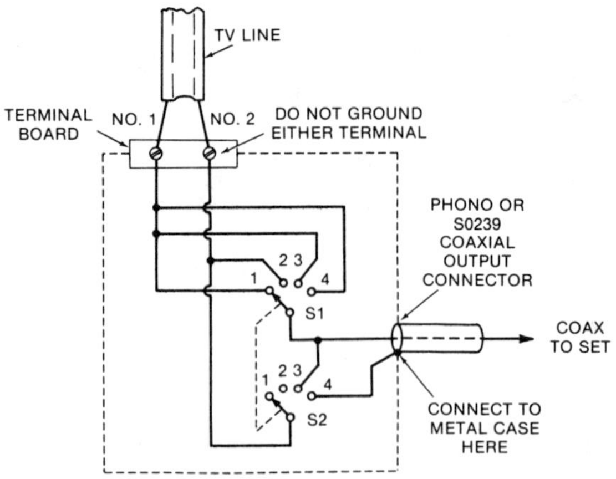

Fig. 5-12. Noll's "flexo" switching.

The basic idea of the flexo connections is introduced in Fig. 5-11. Notice that there are four ways that the two ends of the TV line can be connected to the receiver. The first is the obvious one. This is a conventional dipole connection. In example B just one dipole lead is connected single-ended to the receiver. In this arrangement the length of the end-fed single wire includes one side of the basic dipole and the length of the transmission line itself. In effect it becomes a longwire antenna. The same applies to example C with the exception that the other lead, which is the opposite side of the dipole, is used. In example D the two leads are tied together and fed single-ended to the receiver input. It approximates a single-wire fed "T" antenna.

The practical way of taking advantage of this capability is to use a two-pole, four-position switch as in Fig. 5-12. Often you may have to settle for a switch that has more positions. This is quite OK because you need only wire four positions into the circuit, neglecting the others. Mount in a small metal box. The two wire ends of the TV line are connected to their separate and isolated inputs to the switch. Do not connect either to the metal case. A short piece of coaxial line links the switch box to the receiver. The two switch sections are labelled S1 and S2. The switch is set to position one. Note that wire 1 of the TV line if attached as shown is connected to the inner conductor of the coaxial line and feeds the receiver input in single-wire fashion by way of switch S1. Switch S2 is inactive on position one. On position two, TV line wire 2 is connected to the inner conductor through switch S1. Again, S2 is not active. These two basic

(A) Front view.

(B) Rear view.

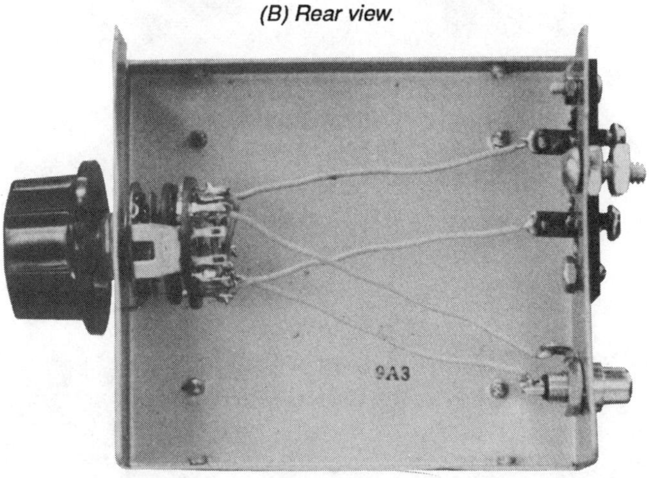

(C) Inside view.

Fig. 5-13. Flexo switch construction.

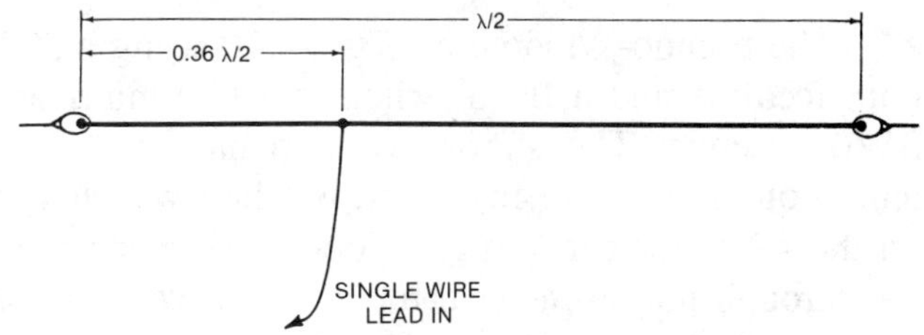

(A) Basic Windom.

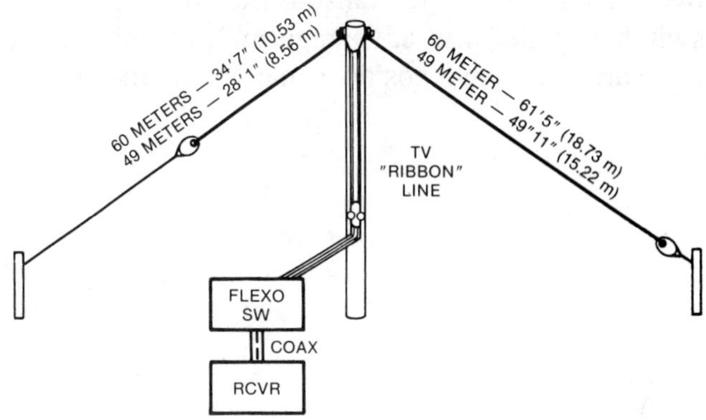

(B) 300-ohm line pseudo-Windom.

Fig. 5-14. Windom SWB antenna.

switch positions provide a simple basic method of switching between two single-wire antennas. On position three, TV line wire 1 is connected to the inner conductor by S1. Now switch S2 becomes active and connects the TV line wire 2 to the inner conductor as well. This sets up the T-antenna configuration. On the final switch position S1 connects TV line wire 1 to the inner conductor of the coaxial line while S2 connects the TV line wire 2 to the braid (via the switch box phono or coaxial connector) of the coaxial line that links the switch box with the receiver. In this arrangement the antenna is being used as a conventional dipole, resonating on 41 meters.

A flexo switch assembly can be built into a small package as pictured in Fig. 5-13. It is a low-cost device and convenient to operate right by your receiver. The small box (Radio Shack 270-251) supports a switch, phono output connector, and a small insulating strip which mounts two screwdriver terminals to which the two wires of the TV line are connected. The switch is a two-pole, six-position type (Radio Shack 275-136) but only the first four positions are employed.

5-8. Pseudo Windom Antenna

At one time the Windom antenna was quite popular among radio amateur operators. Its basic form is shown in Fig. 5-14A. It is an off-center fed antenna and it operates quite effectively as a multiband antenna on those frequencies which are harmonically related to its fundamental length. When a good match is obtained by trimming the length of the feeder, the antenna performs well on a number of bands.

The pseudo-Windom of Fig. 5-14B using a 300-ohm feedline and a flexo switch is a fine multiband SWB antenna. The shorter version has an overall length of 78 ft (23.77 m) which is a half wavelength on the 49-meter band. It is a good performer on the 19- through the 49-meter bands and is acceptable on the 16- and 60-meter bands. Results vary on the 75-, 90-, and 120-meter tropical bands with relation to the overall length of the transmission line. The flexo switch is a definite advantage and, on most bands there are one or two positions that deliver the best signal. As in other antennas, the direction of signal arrival influences the best switch setting.

When erected in inverted-V fashion there is some directivity off the end of the longer leg. In general the antenna is a good omnidirectional performer.

If space is available, you may wish to cut the Windom antenna as a half-wave antenna on the 60-meter band. The dimensions are given in Fig. 5-14. This configuration brings up the performance on the 60-meter band and also is a help on the remaining tropical bands.

Chapter 6

Directional Longwire and Beam Antennas

THE directional (beam) antenna is an arrangement of one or several antenna elements that display maximum sensitivity toward a specific target area when positioned correctly. The basic directional antenna is a single long wire having an electrical length that is a full wavelength or more. However, a 3/4 wavelength wire (Fig. 6-1A) is quite directional off the ends in SWB receiving application. At medium and low frequencies even a quarter wavelength wire (Fig. 6-1B) is quite directional and is often referred to as a "longwire" simply because it takes a long length of wire to obtain an electrical quarter wavelength at these low frequencies. A beverage antenna (Fig. 6-1C) is usually one wavelength long or more and is mounted 10-ft (3.04-m) or less above ground. It is very directional off its far end.

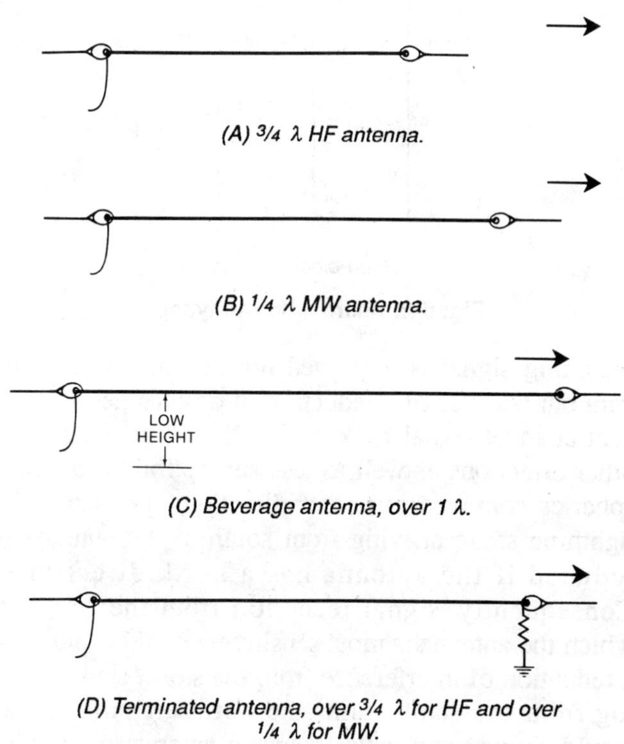

(A) 3/4 λ HF antenna.

(B) 1/4 λ MW antenna.

(C) Beverage antenna, over 1 λ.

(D) Terminated antenna, over 3/4 λ for HF and over 1/4 λ for MW.

Fig. 6-1. The longwire end-fed antenna.

6-1. Introduction

Directional receiving antennas for SWB application also have acceptable sensitivity in other directions because of their relatively low practical height and the resultant blending of their directional patterns into a more omnidirectional configuration. However, as in all directional antenna systems, the improvement in reception in any one direction must be accompanied with a reduced reception capability in other directions.

An additional improvement in the performance of directional longwires can be obtained by using a resistive termination at the far end as in Fig. 6-10. The actual gain improvement in the sensitive direction is limited or none at all. However, the response to an

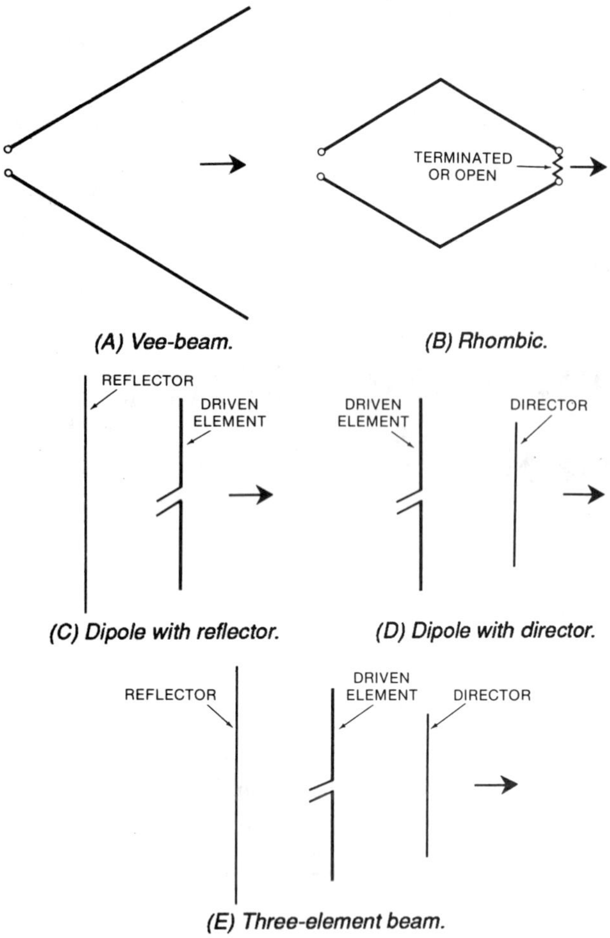

Fig. 6-2. Beam antenna types.

incoming signal is improved not because of antenna gain but because of a reduction of on-channel or adjacent channel signal background "hash" arriving from other directions as well as less susceptibility to atmospherics coming from other directions. For example, lightning static arriving from south of the antenna is reduced if the antenna has a north directivity. Consequently, signal reception from the north to which the antenna is most sensitive would be aided by a reduction of interference from the storm static arriving from the south. Thus the north-arriving signal would sound even better than the antenna directivity pattern would suggest.

Two or more antenna wires can be erected in a vee (Fig. 6-2A) or rhombic (Fig. 6-2B) configuration to build up gain. A good limited-space vee-beam is very possible, especially for the higher-frequency SWB bands.

Dipole antennas with a parasitic reflector (Fig. 6-2C) or director (Fig. 6-2D) are fine gain antennas for shortwave broadcast reception. The inverted type of beam configuration is practical, low cost, and does reasonably well as compared to a strictly horizontal beam when you wish to concentrate on a specific receiving direction. An uncomplicated three-element beam (Fig. 6-2E) provides further improvement. The latter is more appropriate for the higher-frequency bands.

Various phased antennas can be set up for one or two band operation on the SWB bands. The three basic horizontal phased configurations are the collinear, broadside, and end-fire (Fig. 6-3). The broadside style is not too practical for the low-frequency SWB bands because of space requirements and the mechanical problem of stacking one above the other. A collinear design requires quite a spread. Since the end-fire arrangement places one element in front of the other it becomes the most practical arrangement. A unidirectional pattern is obtained with $\lambda/4$ spacing and a bidirectional pattern with $\lambda/2$ spacing.

Vertical antenna elements can be phased as shown in Fig. 6-4. Again, the collinear is impractical for SWB receiving because of the required height. However both end-fire and broadside arrangements are reasonable designs for some backyard builders.

6-2. Slant Wires: Terminated and Unterminated

The quarter-wave sloper antenna was covered in Chapter 5. The slant longwire antenna of Fig. 6-5 provides quarter-wave operation on the 120-meter band for tropical band enthusiasts. The exact 101-ft (30.78-m) length also corresponds to 3/4 wavelength operation on 41 meters and gives a boost to reception on this band. This length also shows a 5/4 wavelength resonance on 25 meters and brings life to the signals on this band. Good directivity, when terminated, also occurs on 49 meters. The terminated longwire has a resistor connected from its far end to ground.

A terminating resistor has a limited effect on the sensitivity of the slant wire in the slope direction. What it does is to suppress the signals that arrive at the back and sides of the slant wire. The same suppression applies to noise and atmospherics. You gain a quieter background so a weak signal coming in from the slope direction is more intelligible.

On the 90- and 120-meter bands the terminating

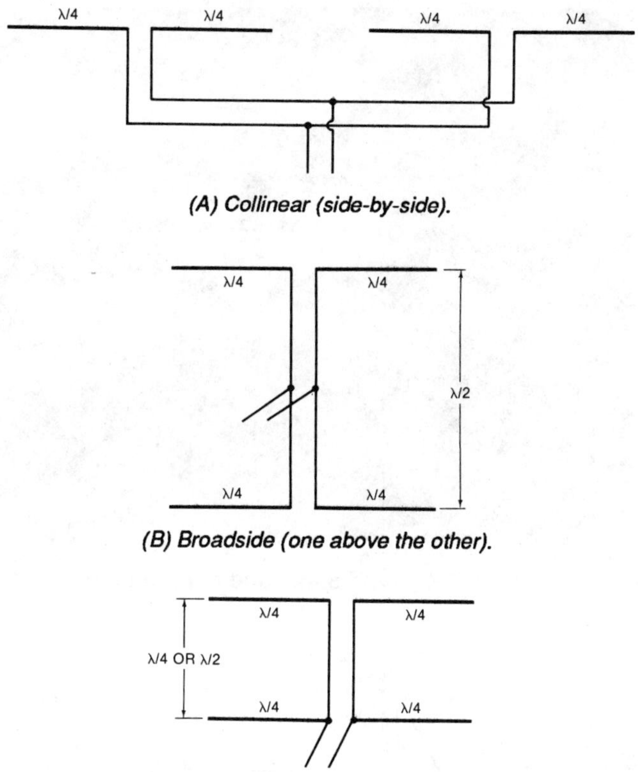

(A) Collinear (side-by-side).

(B) Broadside (one above the other).

(C) End-fire (one in front of the other in the same horizontal plane).

Fig. 6-3. Horizontal phased dipoles.

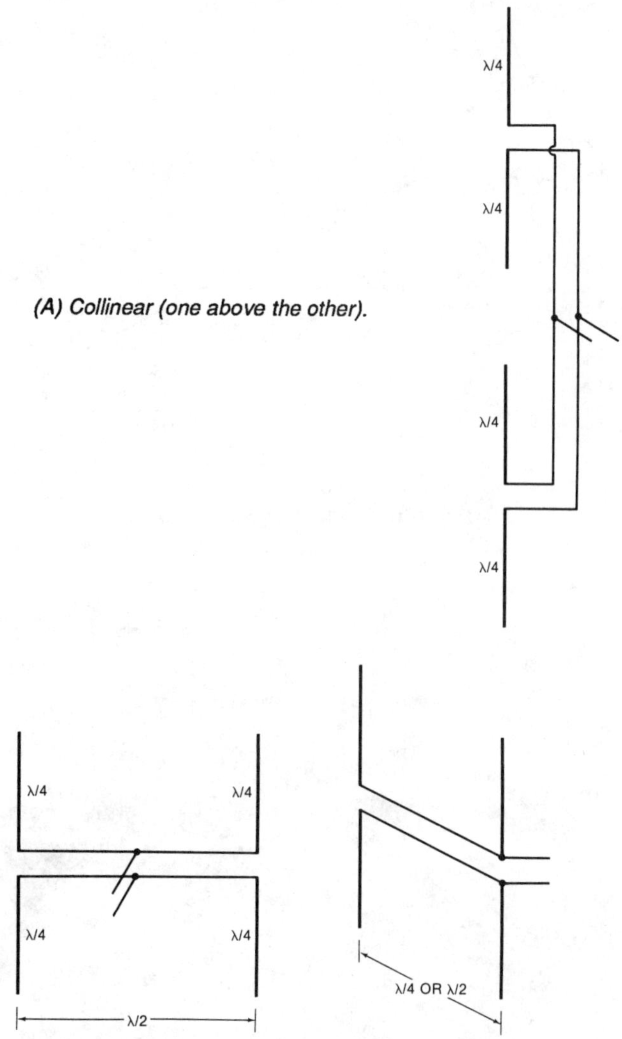

(A) Collinear (one above the other).

(B) Broadside (side-by-side).

(C) End-fire (one in front of the other).

Fig. 6-4. Vertical phased dipoles.

resistor has much less influence and can even reduce signal pickup in the slope direction when used. This is because the wire length itself is short in comparison with a wavelength at these frequencies and the sloper no longer responds as a true longwire antenna. However there still is some limited improvement in the back rejection. Nevertheless, there is acceptable pickup from this antenna in comparison to many antennas that are in use today for tropical band reception despite its limited 101-ft (30.78-m) length.

A terminating resistor can be made from a one watt composition resistor mounted on a small piece of plastic. A suitable value is 220 ohms. A high-power but similar value terminating resistance (Fig. 6-6) is made from three wirewound noninductive 750-ohm resistors connected in parallel as pictured in Fig. 6-6. This trio was operated often on the radio amateur bands using a 200-watt PEP sideband transmitter. A 5-ft (1.52-m) low-resistance ground rod is used with the antenna as shown in Fig. 6-7. Its diameter is 3/8 inch (0.95 cm). The ground rod is driven about 4 ft into the ground. The terminating resistor is connected at the end of the antenna wire and a length of 14-gauge wire links the resistor with the terminal attached to the ground rod.

6-3. A Simple Vee-Beam for 16 Meters Using the All-Band Flexo

A good performing and relatively short vee-beam wire antenna that will fit a small lot can be constructed for any one of the higher-frequency shortwave bands. For example a vee-beam with 3/4-wavelength wires on 16 meters requires leg lengths of only 41 ft (12.5 m) (Fig. 6-8). If you wish to operate this anten-

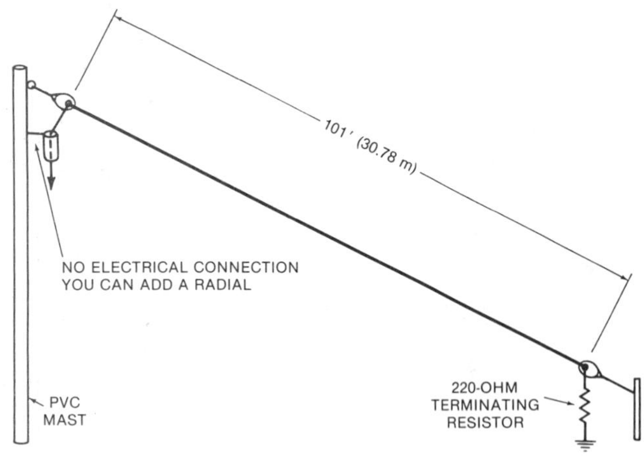

Fig. 6-5. Slant longwire antenna.

Fig. 6-6. Terminating resistors attached to end of slant wire.

na on a lower-frequency band, choose the leg lengths indicated in Table 5-1. For a 3/4 wavelength vee-beam use an included angle of approximately 110°. The installation shown used a 20-ft mast with the wire ends of the beam elevated 10 ft (3.05 m) above ground. In my arrangement the legs were so positioned that with little trouble the beam could be swung around toward Europe in the morning and toward Australia in the evening. This gave me a good idea of how well the beam performed as it was checked against a reference antenna. The performance was also very fine on 19 meters despite the approximate 2.5-MHz frequency separation of the two bands.

Fig. 6-7. Ground stake and connection.

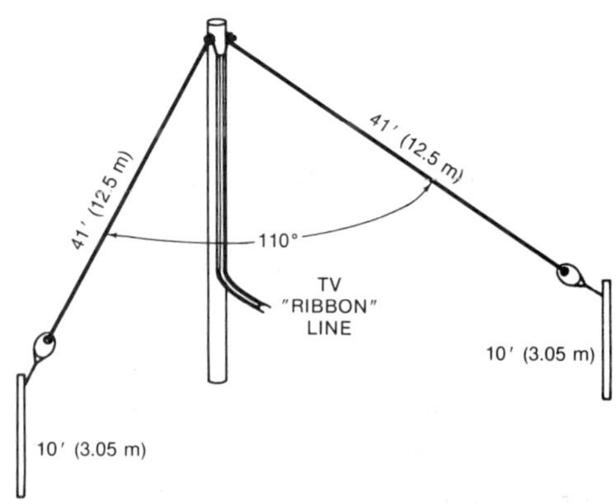

Fig. 6-8. Sloping vee-beam for 16 meters.

The story doesn't end here. On the 49-meter band the antenna functions much as a half-wave dipole and acceptable results are obtained on both the 41- and 60-meter bands. The use of a flexo switch permits good all-band operation. There is usually one or two positions on each band that are better than the other two. On occasion the angle of signal arrival determines the most favorable switch position.

6-4. Inverted Dipole With Single Parasitic Element

Good gain and directivity can be obtained with the use of parasitic elements. (A parasitic element is an element placed in front of or behind the driven ele-

(A) Dipole with reflector.

(B) Dipole with director.

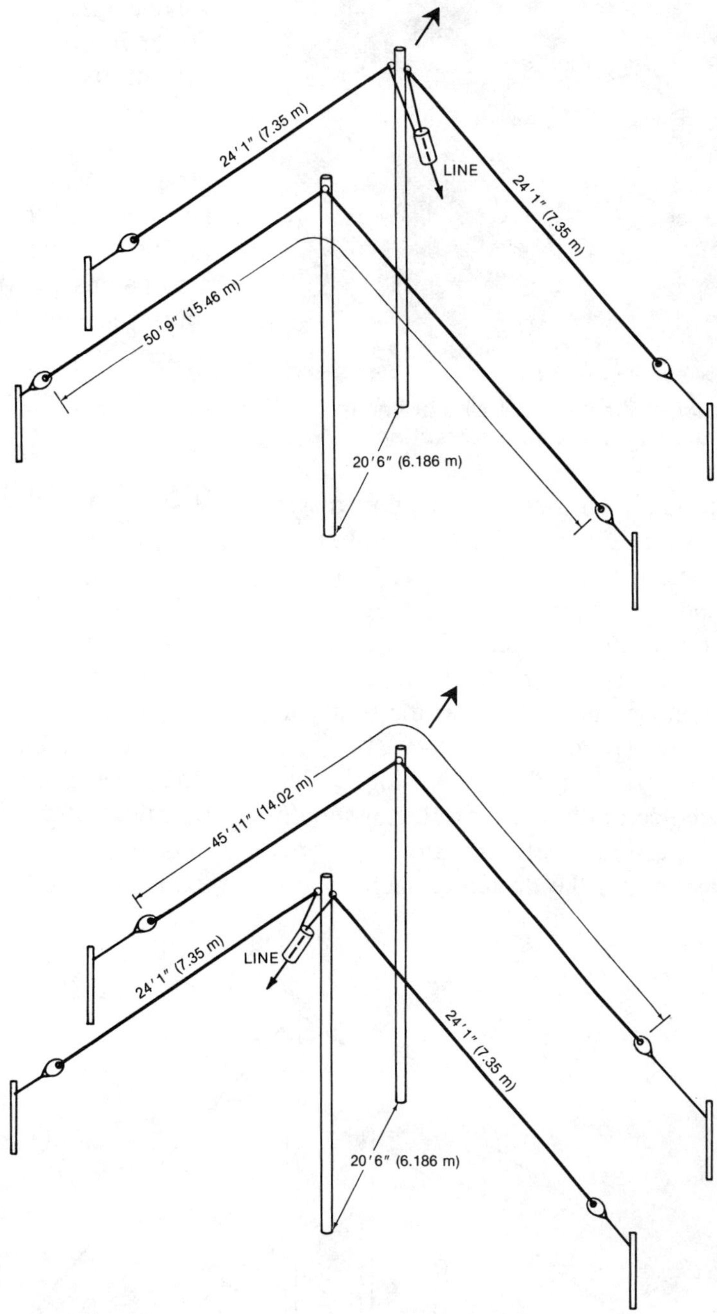

Fig. 6-9. Inverted dipole with parasitic element for 31 meters.

ment but has no connection to the transmission line.) If your shortwave listening includes some single-band specialization such an antenna may be attractive. Such an antenna provides good gain in a favored direction along with reduced side and rear pickup. However, the size of such an array is a problem along with relatively high cost. A boom-type of mount with driven and parasitic elements supported on a single cross-piece is only practical for the highest-frequency bands: 13, 16, and 19 meters. Beams of this type are covered in Chapters 10 and 15. On the lower-frequency bands this antenna requires either four or six masts, depending on the number of elements, and requires considerable mounting space.

A practical alternative is the use of the inverted-V style dipole and parasitic element. There is some

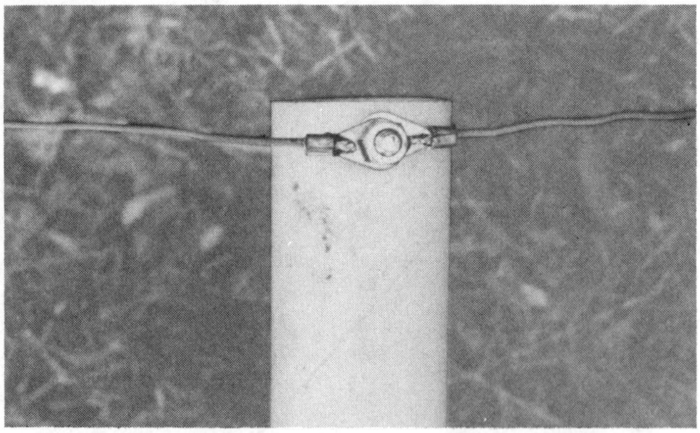

Fig. 6-10. Center of the parasitic wire bolted to the top of the PVC mast before erection.

decrease in performance but only two support masts are required as shown in Fig. 6-9. The drop-off in performance can be held to a reasonable value by keeping the element ends as high as possible. You can do so by stretching out each of the end wires as far as possible with a considerable length of support rope between insulator and ground stake. Do the best you can with your available space.

The dimensions given in Fig. 6-9 are for the 31-meter band. The choice of director or reflector should be based on your particular site situation. Spacing between the parasitic and the driven element is 0.2 wavelength which provides a reasonable match to a 50- or 70-ohm coax line.

In my installation two 20-ft (6.1-m) PVC pipes are used to support the antenna wires. A dipole-to-coaxial line connector is attached to the top of one mast for the driven element. A simple bolt/nut terminal supported the center of the parasitic element at the top of the second mast as shown in Fig. 6-10. The dimensions for the antennas are given in the third, fourth, seventh, and eleventh columns of the SWB dimension tables of Chapter 11 for 31 meters and for other bands as well.

6-5. Three-Element Beam with Reflector and Director

Additional gain and directivity is possible by using both a reflector and a director (Fig. 6-11). In this scheme it is usually necessary to use a somewhat longer driven element. Start with the value given for the beam driven dipole in column eight of the dimension table in Chapter 11. This three-element 31-meter beam did very well receiving those "down under" signals from Australia and also the signal from New Zealand on 9620 kHz.

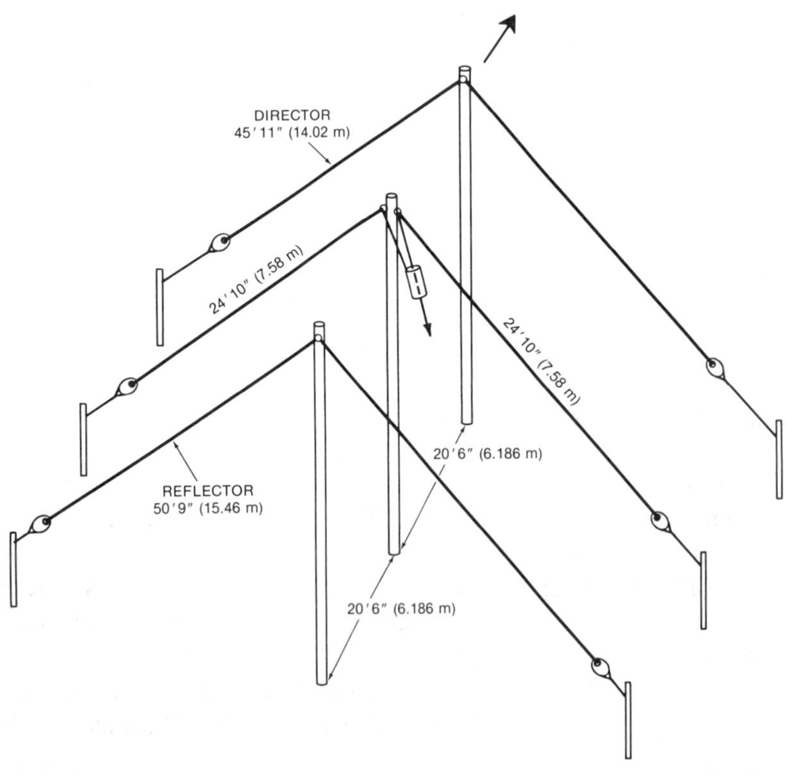

Fig. 6-11. A 31-meter three-element beam.

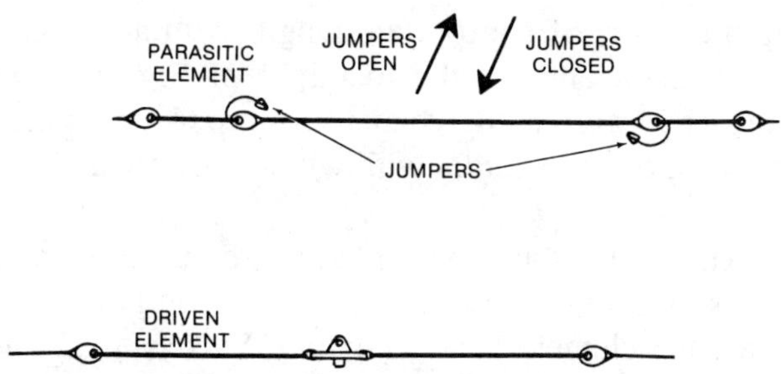

Fig. 6-12. Use of jumpers to change beam directivity.

6-6. Uni-Bidirectional Beam

The beam is not a widely known receiving antenna despite its gain and improved noise-interference pickup. In many locations, because of the intervening oceans, the bulk of your listening may be done in one or two general directions. So why not favor the direction(s) in which you do the most listening? My own location is an example, the preponderance of signals arrive over an arc that extends from just a bit west of north to a bit south of east. A beam aimed northeast does an excellent job. Two- or three-element beams do well despite a mounting height of no more than 20 ft.

Some of my preferred listening is toward the west and southwest. A beam aimed southwest does well. As a matter of fact this second direction is a complement (180° bearing change) of the northeast direction. It is important to understand that a beam using parasitic elements has a back lobe which permits some reception in the complementary direction. It is the side pickup which is reduced when using a beam antenna. If you prefer the ultimate in beam reception this topic shows how one beam can be changed over quite conveniently to receive optimum signal from either of two complementary directions.

An alternative to the problem is the use of two antennas—a vertical for omnidirectional reception, and a beam for some favorite direction. Two transmission lines and a switch give you that versatility.

Isn't a beam a single-band antenna? Quite true, but it can also be used on other bands. Two antennas can solve this problem and permit you to emphasize your favorite band and have reasonably good all-band reception as well.

The basic add-on idea is demonstrated in the two-element beam of Fig. 6-12. Note the length of the parasitic element when the jumpers are not connected is shorter than the driven element. Consequently it acts as a director in a line between the driven element

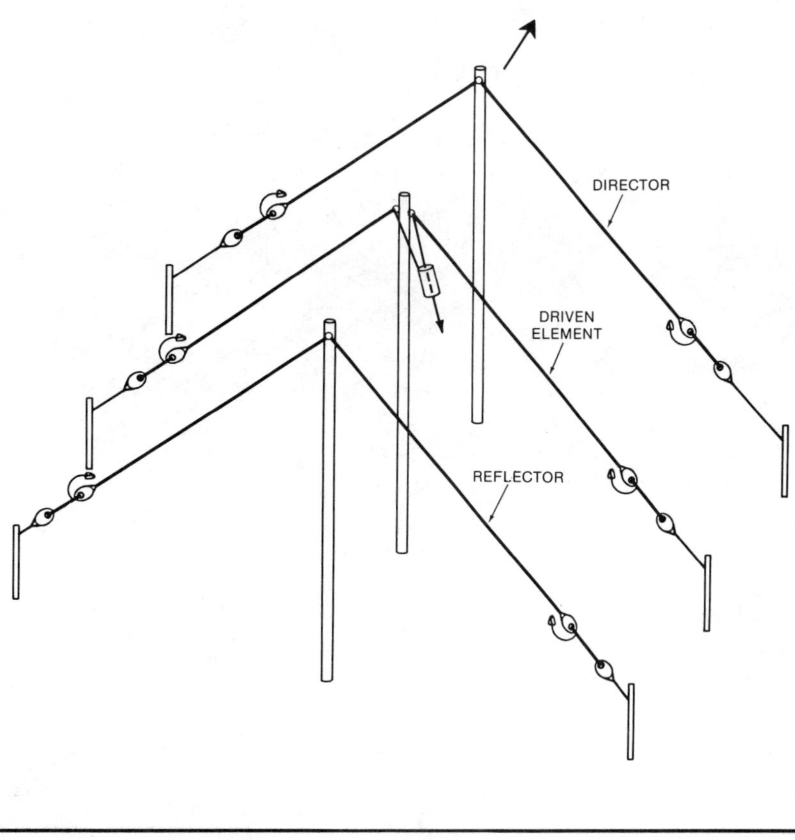

Fig. 6-13. Use of add-ons to permit two-band operation of a three-element beam.

and the parasitic director. When two short add-ons of proper length are jumped into operation the parasitic element becomes longer than the driven one, and acts as a reflector. Now the beam directivity is shifted 180° and points in a line between the parasitic reflector and the driven element.

If add-ons are to be used the most convenient beam configuration is the inverted-V style. The element ends are near ground level and can be easily released one at a time in such a matter that you can gain access to the jumper from ground level or with the aid of a short stepladder. Each jumper is a short length of wire of appropriate length with an alligator clip on each end. In the example of Fig. 6-12 the reflector is activated when the jumpers are placed across the two insulators with which they are associated.

The same changeover plan can be used for making a single beam installation operate on two bands. For a three-element beam (Fig. 6-13) six such add-ons are required for two-band operation. The antenna is originally cut for operation on the higher-frequency bands. Then the six add-ons, when jumped into operation, permit operation on the lower-frequency band.

Chapter 7

Special HF Antennas

THIS chapter opens with a popular VHF/UHF ground plane antenna consisting of a vertical element and a set of fanned-out resonant radials. This style antenna can be adapted for use on the higher-frequency SWB bands and is a good performer. It is easy to construct when using a PVC mast. Phased arrays are also presented including 90° phased horizontal elements which provide a broad unidirectional pattern. Also included in this chapter are verticals fed in phase or out-of-phase.

7-1. Introduction

Full-wave closed-loop antennas are introduced for the first time in this chapter. In this arrangement a full wavelength of wire is folded back upon itself. The transmission line is attached where the two ends of the wire are about to meet as shown in Fig. 7-1. The earliest and most common full-wave configuration is the quad (Fig. 7-1A) which, when operated with a reflector, is a popular ham radio antenna because of its fine performance albeit its erection difficulties and less than good track record for staying up and in shape in bad weather areas. The quad is also used in VHF/UHF application. The small antenna size at these frequencies permits a stable construction.

The delta loop (Fig. 7-1B) is another version of the quad, as are the diamond (Fig. 7-1C) and the circle (Fig. 7-1D) used occasionally in VHF/UHF application. The triangle (Fig. 7-1E) is the most physically stable configuration of the group and provides good performance. Such a wire triangle can be supported from ground level with guys or low posts. Only a single high support mast is needed to sustain a rigid structure. Of the various configurations the triangle is by far the most convenient for use at low frequencies. Several triangle antennas are detailed in this chapter.

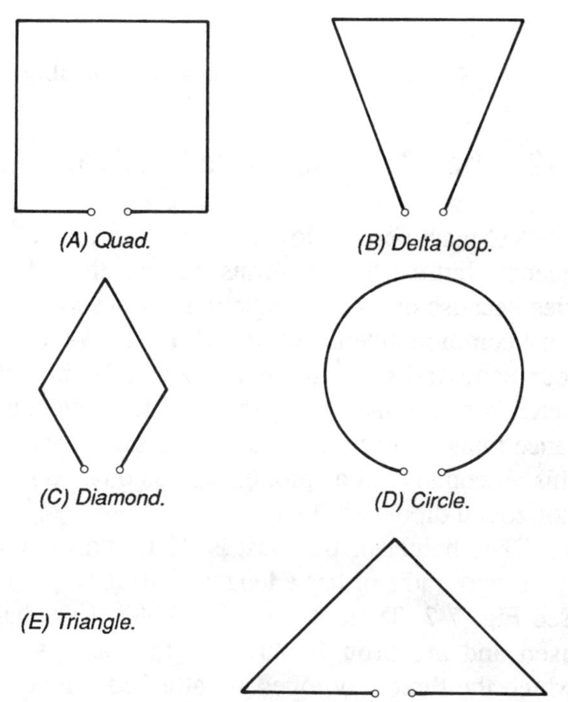

Fig. 7-1. Full wavelength configurations.

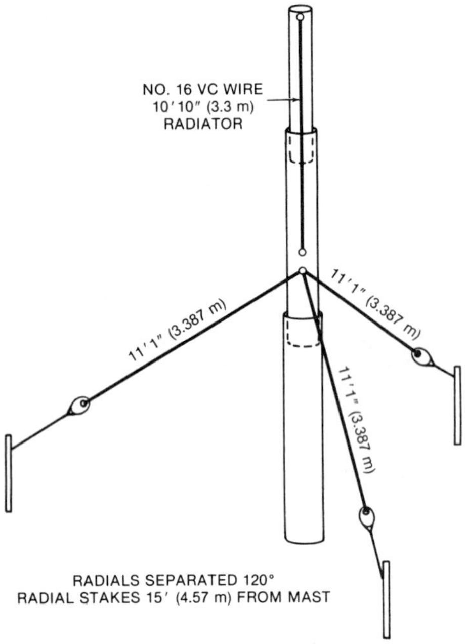

Fig. 7-2. A 13-meter elevated vertical.

7-2. The 13-Meter Wire Vertical

A PVC mast permits low cost erection of a high-frequency vertical that performs well on these frequencies because of its low angle vertical response. It is a very common antenna on the 10-, and 15-meter amateur bands all over the world. Actually, it performs better for long haul reception than for a medium distance range. Sometimes the deepness of a fast fade on this antenna is not as pronounced as on a comparable horizontal dipole antenna.

The height of the mast is 24 ft and the quarter-wave wire radiator has a length of 10 ft 10 in (3.3 m). See Fig. 7-2. Three 11 ft 1 in (3.387 m) radials are used and are brought down to the same stakes to which the three guy ropes are attached. Proper radial length for elevated HF types may be shorter than values given in the radial column of dimension tables. A bolt/nut combination at the top of the mast supports the top of the vertical. Bolt/nut terminals support the bottom of the radiator as well as the tops of the three radials (Fig. 7-3).

An advantage of a PVC mast is that insulators are not needed to support the vertical radiator and the electrical interactions that are often present with a metallic mast do not exist. Another advantage of the plastic mast is that the same mast can support other antennas without mast interaction. The rope shown in Fig. 7-3 is a lanyard arrangement that can be used to raise and lower slopers up and down the same mast with ease.

Fig. 7-3. PVC mast showing wire vertical, three resonant radials, and two bolt/nut terminals.

7-3. The 31-Meter Triangle Antenna

The triangle is the stable member of the full-wave-length loop configuration. In HF service it performs well, is easy to construct, and weathers well. It is called a triangle because, unlike the delta loop, it does not imply that all sides are the same length. It is usually fed at the center of the triangle base leg, but it can also be corner fed. The base of the triangle can be significantly longer than the sides before full-wave loop performance deteriorates a serious amount.

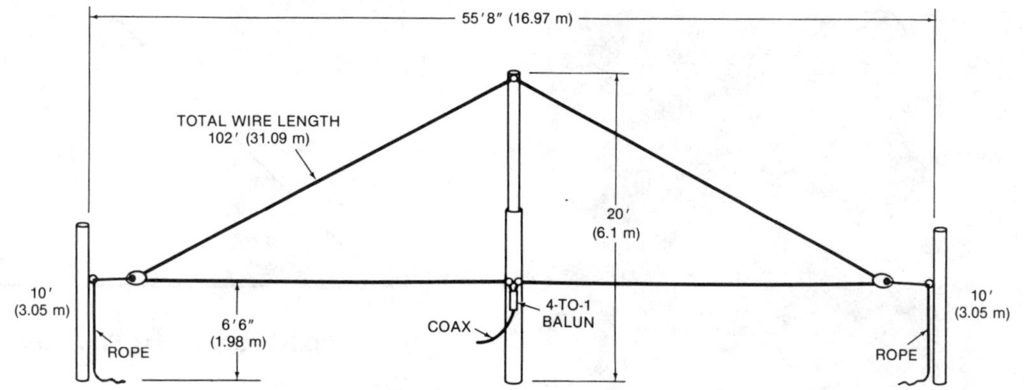

Fig. 7-4. A 31-meter triangle.

The triangle configuration permits you to erect a good multiband antenna, peaked on the band for which it is cut, using only a single high mast and two low support points as shown in Fig. 7-4. On the band to which it is cut, its horizontal sensitivity pattern is much the same as a dipole. It provides a slight amount of signal gain over a dipole, however. Another DX feature of the triangle that is of importance to the DXer is that despite a low mounting position it maintains good low angle vertical directivity. For those who live in a noisy location it also provides a low background noise configuration. Maximum directivity is broadside to the plane of the triangle.

The construction arrangement and dimensions for the 31-meter triangle are given in Fig. 7-4. The total length of the wire is about 102 ft (31.09 m). The full wavelength figures are given in Column 8 of the dimension tables in Chapter 11. In the construction procedure two 51-ft (15.54-m) lengths of No. 14 insulated wire are cut. One end of each wire is bared and a solder ring attached. The two ends are slipped over a bolt/nut combination at the top of the mast before it is erected. The other end of each wire is passed through one eye of an insulator and then bared for attachment of a second solder ring. These will be attached to bolt/nut terminals mounted on the main mast about 6 ft 6 in (1.98 m) above ground. I attached a single guy rope near the top of the main mast.

Next drive the fence posts that will support the main mast and the two lower masts into the ground in a line perpendicular to the preferred directions of maximum sensitivity. (My installation was on a line northwest-southeast and the triangle displayed top sensitivity toward the northeast and the southwest.)

Fig. 7-5. Each half-wave segment is passed through one eye of an individual insulator. A rope passes through the other eye and holds the triangle corner to a low support point.

Attach a pair of support ropes to the opposite ends of the insulators and drop through the two eye-ring bolts mounted 6 ft 6 in (1.98 m) up on the two short masts or other low support points.

Now the main mast can be raised. Next the two wire ends are connected to the main mast antenna terminals. The two end ropes are pulled so that the wires pull out into a triangle (Fig. 7-5). You may have to adjust each end and the single guy rope alternately until you have the mast standing perfectly upright. In shaping the triangle a short ladder helps you to adjust the amount of wire on each side of each end insulator. Usually insulated wire gives you no problem in shaping a good triangle. It is not as easy when using bare wire that likes to twist. You need a bit more patience.

Fig. 7-6. Ends of the antenna wires attach to terminals mounted about 6 ft 6 in (1.98 m) high on mast. Also shown are the balun and antenna meter.

The antenna feedpoint resistance varies with the length of the triangle base and, the height that it is above ground in terms of wavelength. In general the higher the base, the higher the antenna resistance, varying between about 50 and 200 ohms. My own installation measured about 180 ohms at 9.8-MHz resonance. A 4-to-1 balun (Fig. 7-6) reduced this value to that needed for an acceptable match to a coaxial line. The antenna bridge is of great value in measuring feedpoint resistance and resonant frequency of the full-wave antenna configuration.

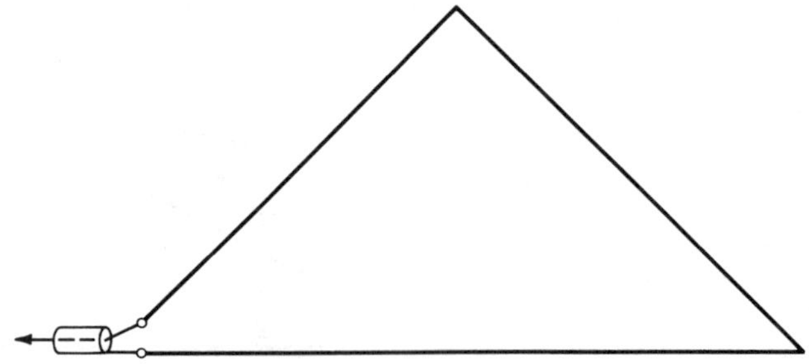

Fig. 7-7. Corner feeding a triangle.

This array turned out to be a good antenna with a bidirectional pattern and somewhat more gain than a reference dipole. The side pickup was somewhat less. Drop was quite significant right along the northwest-southwest line.

The performance was also quite good on the SWB bands lower in frequency than the 31-meter band. The results were poor on the higher-frequency bands except for 13 meters. Of course you can build your triangle for any band using the dimensions given in the dimension tables in Chapter 11.

The triangle can also be corner fed as shown in Fig. 7-7. The unusual feature of this arrangement is its lower feedpoint resistance and the fact that a direct connection to a coaxial line can be made using the simple dipole-to-coaxial connector arrangement of Fig. 7-8. No balun is needed. The performance is the same or slightly better on the long haul signals when it is corner fed.

Note in Fig. 7-4 that the overall length of space that the triangle occupies is less than that required by a similar dipole. Furthermore the ground space occupied by a two-element triangle is even less than that required by other two-element horizontal beams.

7-4. Triangle and Reflector

A two-element triangle antenna (Fig. 7-9) occupies less space than a similar driven dipole and reflector combination because the optimum spacing between the triangle and its triangular reflector is only 0.12 wavelength. Notice the spacing approximates only 12

Fig. 7-8. Corner dipole-to-coax connector.

ft (3.66 m) in the case of the 31-meter two-element beam. In terms of practical results most experts rate the full-wavelength two-element loop beam somewhere between the results obtained by two-element and three-element parasitic horizontal beams. This performance is possibly related to the good low-angle directivity. I oriented mine toward Australia and was rewarded with excellent 9580–9610–9770 kHz morning signals along with a readable New Zealand signal on 9620 kHz.

The reflector wire was made approximately 3% longer than the wire of the driven triangle. It was cut in half so the exact center of the triangle reflector could be located at the very center of the high mast. The top of the reflector center mast was made about 1 ft 2 in (35.6 cm) higher than the driven one with the addition of a short 1-in (2.54-cm) ID length PVC pipe telescoped into the top of the mast. Also, the single base connection terminal for the triangle reflector was dropped the same amount lower than the two antenna terminals mounted on the driven triangle mast.

The erection procedure was the same. However the connection at the base of the triangle reflector mast was completed, forming a closed reflector. When a reflector is used the driven triangle impedance falls to a value that permits a direct connection to a coaxial transmission line.

7-5. Special 90° Phased Unidirectional Dipoles

Two properly spaced and phased dipoles produce a useful, broadly directional antenna pattern. Such an antenna can produce directivity over an arc of nearly 180°. If you wish such a pattern on some favorite band in a certain direction, the 90° phased dipoles are worth consideration. There is some pickup in other directions; actually the assembly is quite an accept-

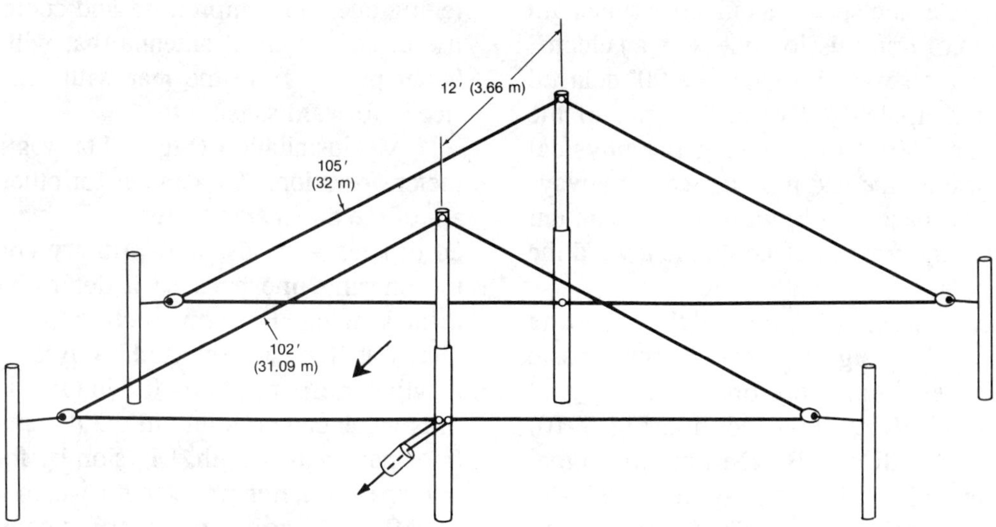

Fig. 7-9. A triangle driven/reflector combination.

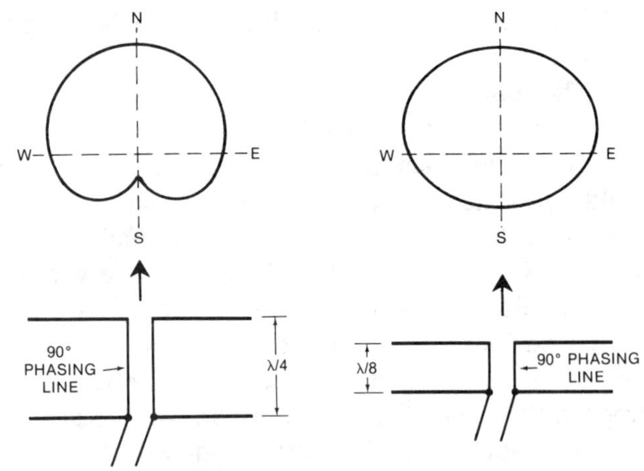

(A) Quarter-wave spacing between dipoles.

(B) Eighth-wave spacing between dipoles.

Fig. 7-10. Patterns for 90-degree phased dipoles.

able multiband antenna even though it gives that special additional edge to your favorite band.

The basic concept of 90° (quarter-wave) phasing is illustrated in Fig. 7-10. When two dipoles are spaced a quarter wave (90°) from each other and the front dipole is fed a quarter wave later than the rear dipole with a λ/4 section of transmission line as in Fig. 7-10A, the sensitivity pattern is the cardioid shape shown. Note the gain that exists over the west-north-east arc when the array is directed north. Of course, the pickup is poor for the southern arc. From my location such a pattern directed approximately northeast includes all of Europe, the Middle East, as well as much of Asia and Africa.

Note the dipoles are spaced a quarter wavelength and that a section of transmission line with an electrical length of a quarter wave provides the 90° delayed feed for the front dipole. The velocity factor of the line must be considered in choosing the physical length of the phasing line and the line will be always shorter than a free-space quarter wave. The maximum directivity is in a line from the feed dipole toward the delayed feed dipole.

A brief consideration of how the antenna responds to an arriving signal gives you some basic idea of just how such an array operates. A signal arriving at the front of the array (north in Fig. 7-10) first excites the front dipole. By the time the signal has arrived at the fed dipole, it has been shifted 90°. However, the arriving signal as it arrives at the second dipole has also been shifted 90°. Consequently the contribution by each dipole at the point where the transmission line is attached is the same and the two signal voltages are in phase. The two signals are additive at the point where the transmission line attaches. Conversely, a signal arriving from the opposite direction first excites the fed dipole. Then, as signal arrives at the front dipole it has been shifted 90° because of the greater distance it travels. In its travel along the phasing line back to the fed dipole it encounters still another 90° shift. As a result the two components at the line terminals are out-of-phase. They are subtractive and the net voltage across the transmission line terminals is small.

The example in Fig. 7-10B is for two dipoles spaced only one-eighth wavelength but fed in a 90° relationship. The resultant pattern is a circle with a flattened lower half. The pick up is higher north of the east-west line and lower over the southern arc but not as pronounced as in Fig. 7-10A. The phasing line remains 90° in electrical length. If the dipole phasing is somewhere between one-eighth wavelength and one-quarter wavelength the resultant pattern would fall somewhere between that of the two examples in Fig. 7-10, permitting a somewhat improved southern arc pickup and accommodating to a simple phasing line arrangement.

The ultimate in performance for the antenna in Fig. 7-10A requires straight-line dipoles, careful matching, precise resonance of each dipole and precise 90° phasing lines, and, preferably, a balun to balance the complete array relative to the feedpoint. However, in building such an array for shortwave listening one can compromise and come up with a very useful and practical antenna that will also have a bit better pickup from the rear with some limited sacrifice in forward sensitivity.

My installation (Fig. 7-11) was planned for 19-meter operation. You can cut for other bands with the aid of the dimension tables in Chapter 11. Two inverted dipoles were used. A primary consideration was the phasing line because it determines the required mast spacing. For construction convenience a 300-ohm flat line was selected. A free-space full wavelength corresponds to 64 ft 4 in (19.6 m) at 15.3 MHz. This value can be found in the dimension table under free-space wavelength. Division by four determines a free-space quarter wavelength which is about 16 ft 1 in (4.9 m). Next it is necessary to consider the velocity factor of the line. To determine the physical length

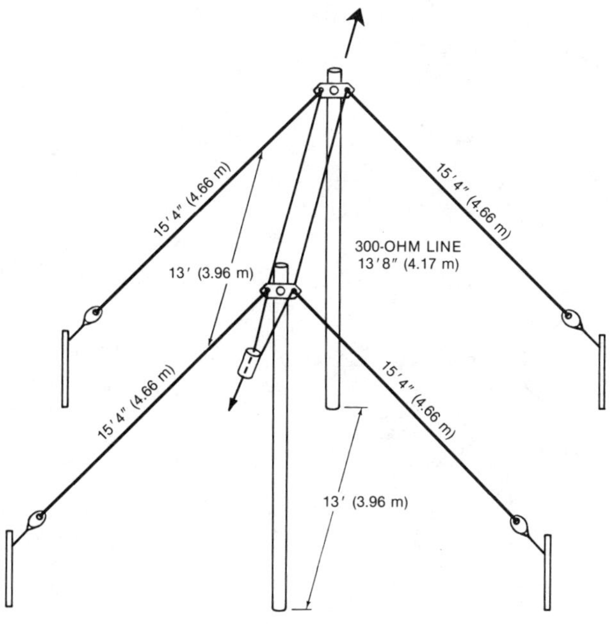

Fig. 7-11. A 19-meter 90-degree fed end-fire array.

of the phasing line needed to present an electrical quarter wavelength on the 19-meter band use a velocity factor (VF) somewhere between 0.82–0.86. Thus, the physical length of phasing line (using 0.85 VF) is:

$$\begin{aligned}\text{Phasing line} &= \text{VF} \times \lambda/4\\ &= 0.85 \times 16.083\\ &= 13.67 \text{ ft}\\ &= 13 \text{ ft } 8 \text{ in } (4.17 \text{ m})\end{aligned}$$

A solder ring is attached to each of the four wire ends of the 300-ohm line to permit attachments to the pair of bolt/nut terminals mounted at the top of each mast.

Fig. 7-12. Two dipoles and 90-degree phasing line between mast tops.

Here each dipole is connected as well. The transmission line must also be attached to the appropriate pair of terminals.

The spacing between the masts is 13 ft (3.96 m) to accommodate this length of phasing line which must link the two dipoles (Fig. 7-12). Such a spacing is a bit less than a quarter wavelength in free space. Hence the antenna's actual pattern will not be as pronounced a cardioid as Fig. 7-10A but not as rounded as the pattern of Fig. 7-10B.

The antenna did a fine job over a wide sweep of land mass when the lines of the two masts were positioned for a bearing about 15° east of true north. The array also did well as a general purpose multiband antenna. I dismantled it with disappointment when space was required for other test antennas. The mast height was about 20 ft (6.1 m) and the dipole ends were approximately 15 ft (4.57 m) above ground.

Chapter 8

Medium- and Long-Wave Antennas

An antenna wire with a physical length that corresponds to an electrical quarter wavelength on the broadcast band is an important objective. When a quarter wavelength wire is open at one end there is a low impedance at the opposite end when that antenna is connected to a receiver input. This is a favorable match for the low-impedance inputs of modern broadcast band (BCB) receivers. The intercepted signal is thus applied efficiently to the receiver input. To be effective at the high-frequency end of the broadcast band the overall length of the antenna wire should be at least 150 ft (45.72 m).

8-1. Quarter-Wave Broadcast Band Antennas

Typical quarter-wave antenna lengths are given in the dimensions of Chapter 11 (Table 11-2). Note that a quarter wavelength corresponds to 146 ft (44.6 m) at 1600 kHz and up to 433 ft (132 m) at 540 kHz at the other end of the band. If you are searching for peak signal interception and maximum signal transfer to the receiver input, the table indicates that a variety of lengths should be employed. However, the power of modern stations and receiver sensitivities permit one to be much more practical in choosing an appropriate antenna length. Nevertheless the wise choice of length is of significance in optimizing the overall performance.

The three typical wire lengths given in Fig. 8-1 favor the high-, mid-, and low-frequency segments of the broadcast band. Any one of these lengths will provide good performance over the entire band but will favor the segment of the spectrum to which the wire length has been cut. If you use a single-wire feed line, that line itself can be considered a part of the overall length of the antenna. For best performance the antenna wire should run in a straight line. However, the antenna works well when it is only approximately straight and even folded at a 90° angle. The antenna can even be made three-sided or, in the extreme case, folded back upon itself very sharply.

Fortunately, a good antenna can be erected in a small space using the fold-around procedure. Several examples are given in Fig. 8-2. The use of three or four PVC masts makes the job easy. If you select a midband length of 240 feet, the erection area required for a three-sided quarter-wave is about 80 ft by 80 ft; a four-sided antenna is only 60 ft by 60 ft. A double loop brings you down to a 40 ft by 40 ft plot or less.

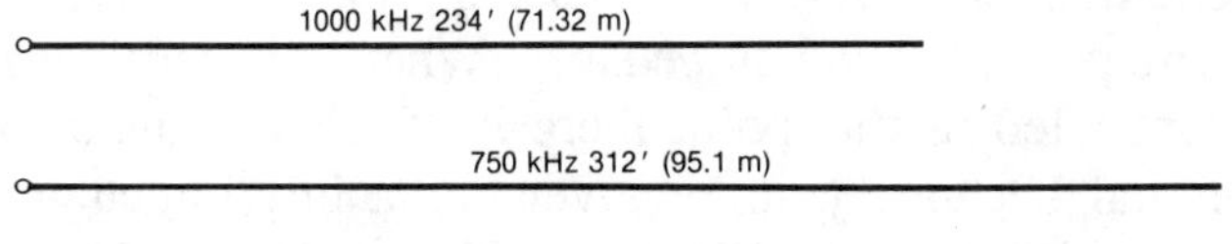

Fig. 8-1. Quarter wavelengths on the broadcast band.

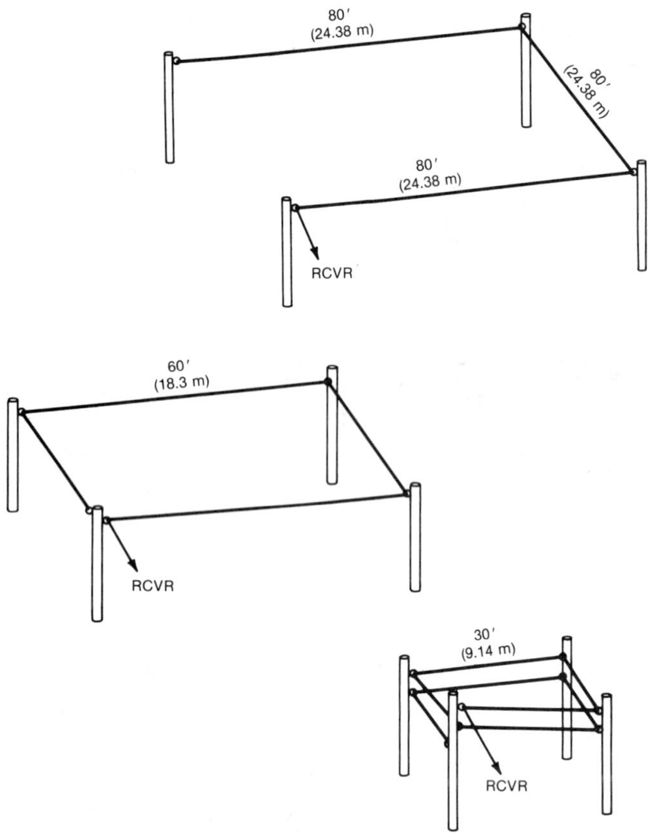

Fig. 8-2. Folded quarter-wave antennas.

The loop need not be square but can be rectangular or some other irregular shape. If listening to the "graveyard stations" (the many low-power local stations that operate 24 hours on 1230, 1240, 1340, 1400, 1450, and 1490 kHz) is your specialty consider an overall length of about 160–180 ft (48.77–54.86 m) for your antenna. Various quarter-wave variations are covered in the topics that follow. Short medium wave/long wave (MW/LW) antennas are presented in the next chapter.

8-2. The 3/4 Wave MW/LW Antenna

The end impedance of an end-fed antenna wire is related to its electrical wavelength. There is always a high impedance at the far end of the antenna wire and one-quarter wavelength from the far end the antenna wire presents a low impedance. When your receiver is connected to this point there is efficient transfer of signal because your receiver has a low-impedance input. Things are different with a half-wavelength antenna. Again, the impedance at the far end is high. However, with a half-wavelength of antenna wire between the far end and the receiver input, the impedance of the far end repeats itself and therefore you are connecting a high-impedance antenna point to the low impedance of your receiver input. Consequently an impedance mismatch exists and the transfer of signal is not as good. This is not to say that no signal is transferred but the amount transferred is not as great as that which occurs when there is a proper match. For lengths between one-quarter and one-half the efficiency of transfer gradually diminishes.

When the antenna wire becomes greater than a half wavelength long the efficiency of transfer gradually rises, reaching a new peak with an overall length of three-quarter wavelength. Thus, one-quarter wavelength from the far end the impedance is low, one-half wavelength from the far end the impedance is high, and three-quarter wavelength from far end the impedance is again low. A simple statement summarizes this fact: an end-fed antenna wire has a low impedance at the receiver end for wire lengths that correspond to an odd-multiple of a quarter wavelength.

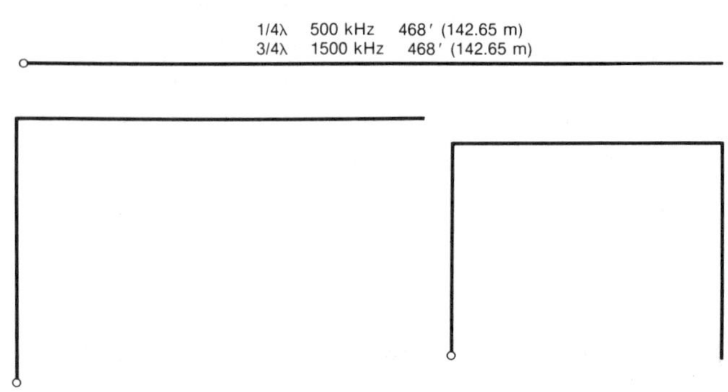

Fig. 8-3. Medium-wave, long-wave longwire antennas.

An interesting antenna is shown in Fig. 8-3. Here the antenna is cut for 500 kHz as a quarter-wavelength. The same length corresponds to three-quarter wavelength on 1500 kHz. This antenna also performs very well on the low frequencies if you are a long-wave fan. In the frequency range around 500 kHz you can copy ship-to-shore CW communications and, as you go lower in frequency, the various airport and marine beacon and weather stations come through. You can also search for foreign LW broadcast stations.

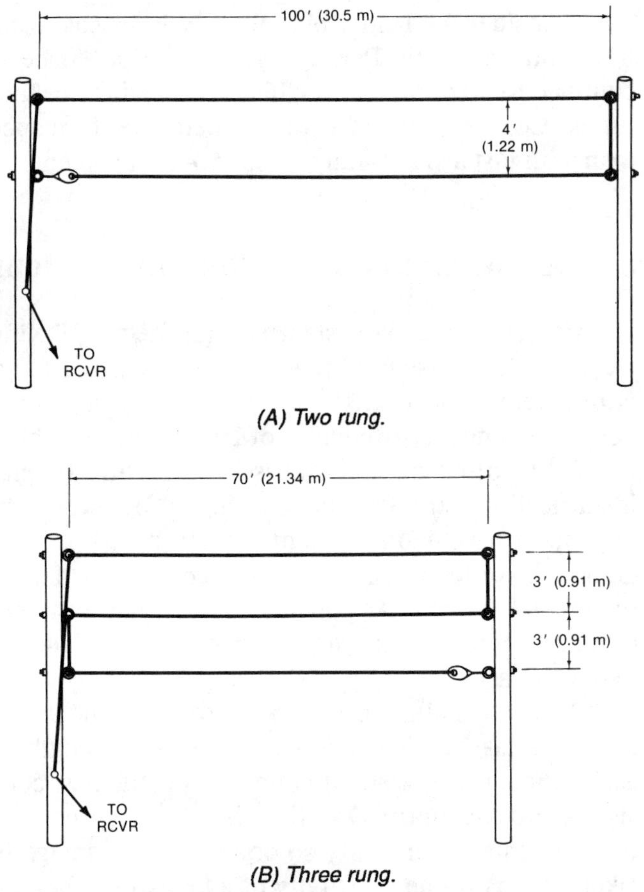

Fig. 8-4. Two and three rung ladder antennas.

8-3. Ladder Antennas

The basic 100-footer (30.5 m) is not the ultimate MW/LW antenna for a given amount of space. Performance can be improved using one of the ladder arrangements shown in Fig. 8-4. Fig. 8-4B occupies less space than the basic 100-footer (30.5 m). Ladder antennas can be erected with very little trouble using PVC piping masts.

Each mast consists of two or three telescoping sections according to the height you desire. For a two-mast section be certain to use a pair of back-up guys at each mast. Additional guying is helpful in erecting a three-section PVC mast. Mast construction detail was given in Chapter 2.

In Fig. 8-4A the two masts are separated by 100 ft (30.5 m). In preparing each mast two additional screw eyebolts are inserted 4 ft (1.22 m) below the top pair. Note how the antenna wire runs up the first mast to the top eye ring and through it to the top eye ring of the second mast. It then goes down 4 ft (1.22 m) and

Table 8-1. Signal Level of Received Ladder Signals Compared to Same Signal Picked Up on 100' (30.5 m) Longwire Antenna

Freq. (kHz)	Two-Rung Ladder	Three-Rung Ladder
560	+3 1/2	+3
610	+1 1/2	+1 1/2
660	+2	+1 1/2
710	+1	+2
770	+3	+2
830	+4	+2
880	+3 1/2	+2 1/2
920	+4	+2 1/2
990	+6 1/2	+5 1/2
1030	+6	+3
1090	+6	+5
1130	+3 1/2	+3
1170	+2 1/2	+3
1210	+ 1/2	0
1260	−1 1/2	0
1320	+1	0
1370	−3 1/2	−2 1/2
1420	+1	−1
1470	−2	− 1/2
1540	+2	+3
1590	−3 1/2	−4 1/2

returns to the lower eyebolt of the first mast where it is looped and taped. Consequently, the overall length of the antenna wire is greater than 200 ft (61 m). There is a decided improvement in gain particularly over the middle- and low-frequency ranges of the broadcast band. Either single-wire or coaxial feed can be run from the antenna wire to the radio room.

If you are cramped for space, use a three-rung ladder as shown in Fig. 8-4B. Mast separation is only 70 ft (21.34 m) and the overall antenna-wire length is greater than 200 ft (61 m). In my antenna the rungs are spaced 3 ft (0.914 m). Although the performance of the three-rung version is poorer than the two-rung type it is still better than a single 100-footer.

Gain figures are given in Table 8-1. Values shown represent S-meter readings above that obtained from a single 100-footer (30.5 m). Note in particular the high gain readings obtained over the midband. An antenna wire over 200-ft (61-m) long resonates as a quarter-wave in this midfrequency spectrum. At the high frequency end of the broadcast band the results are more spotty, the response going up and down, as influenced by the direction of signal arrival. However, over the mid- and low-spectrum the gain figures are always high. This observation also applies to the three-ladder arrangement. Signal values are lower than for the two-ladder plan but still greater than that of the basic 100-footer (30.5 m) except at the very high-frequency end of the broadcast band.

The figures in Table 8-1 are reasonably valid because of the method of measurement. In the testing of broadcast band receiving antennas I use a 230-ft (70.1-m) reference antenna that has been erected in the attic crawl space to keep it unchanged and well separated from the antennas to be tested. This is a ladder antenna that has been strung among the attic beams of the house. In preparing Table 8-1 the 100-footer (30.5 m) was put into position first and a set of figures obtained as compared to the attic reference antenna. Next, the 2-rung ladder antenna was put in position in exactly the same space simply by adding more antenna wire to the 100-footer. Then a second set of figures was obtained. The figures tabulated in the table were obtained by subtracting the decibel (dB) figures plotted for the two ladder from those figures obtained previously for the basic 100-footer (30.5 m).

Next one of the masts was moved in a straight line 30 feet nearer to the first mast. Then the three-rung antenna was checked out and a new set of readings taken and compared with the reference. These new figures were then compared with the figures obtained previously for the 100-footer (30.5 m). The different values are shown in Table 8-1. Thus, the figures you see in the table are really a direct comparison with the performance of a 100-footer (30.5 m).

I found this comparison method the only reliable way to check out the performance of a broadcast band antenna. You can't always obtain a true picture by using separate antenna positions and erections for the two antennas you wish to compare. Also, taking one down and replacing it with another in the same position without having an intermediate reference (such as the attic reference antenna used here) does not provide a true picture because of the many variables of propagation and site location.

You might query, "What is the importance of additional antenna gain when receivers are so sensitive and stations so powerful?" One need only ask, "What about the weak signals?" There are more weak signals than strong ones and it is the weaker ones you have the most difficulty identifying. An improvement in a signal level of 2 or 3 dB can mean the difference between identification and no identification. At other times you may enjoy more comfortable listening from some program that you can only receive from a distant or weak regional or local station.

Consider the ladder plan for any broadcast band horizontal antenna. Perhaps your erection space is confined to 40 or 50 feet. In this case use three or four rungs. Certainly you will obtain better performance than with just a single-rung 40- or 50- foot antenna.

8-4. A Switchable BCB Antenna System

A switchable antenna system is a big help in DXing weak and long-distance broadcast signals. Since longwire antennas have differing directional characteristics that change as a function of length, the switching capability often permits "crossover" IDing (signal identification) on a frequency occupied by more than one station. On one antenna station No. 1 will dominate while station No. 2 will come in better on the second antenna. As you switch back and forth two or more stations on the same frequency can be identified a bit easier.

In my installation separate, low-end and high-end antennas were erected with most of the length of each running due west. Of course this is the best consistent cross-country DX direction from an eastern Pennsylvania location. By so doing best advantage is taken of the off-end directivity of a longwire. The two radio room ends of the antennas terminate at the two inner conductors of a two-position coaxial switch (Fig. 8-5). A short single wire connects to the output of the coaxial switch to the single-ended input of the receiver. Any good single-pole double-throw switch will do. The coaxial switch was already in position here and it was convenient to make the connections shown. An average 65-ft (19.8-m) separation was maintained as the wires were supported by PVC pipes and tree limbs. I use a good water-pipe ground.

The performance is condensed in Table 8-2. Note that the band was subdivided into five receive groups. For each segment stations were located in four compass directions within thirty degrees. By switching between the two longwires it was possible to record the S-unit difference in signal pickup. The final column shows the S-unit ratio between the long- and short-wire readings. Note in segment one that the longwire was completely dominant in all directions. The second segment shows the performance of the short wire catching up to that of the longwire with the exception of the westerly direction. Here the west directivity of the longer wire becomes apparent.

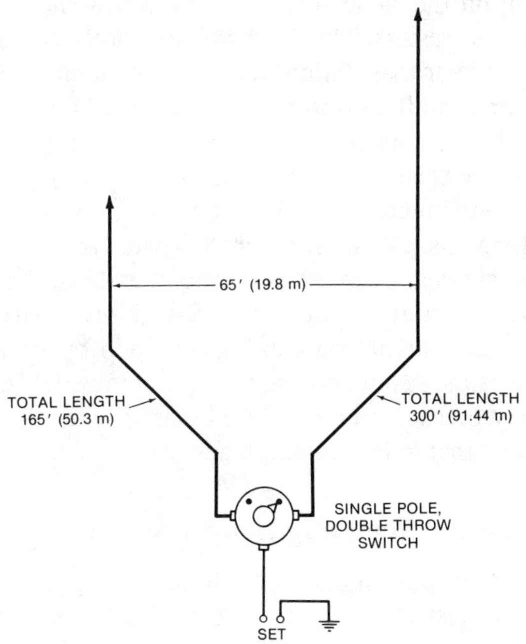

Fig. 8-5. Separate longwires for high- and low-frequency ends of MW band.

Section three is "touch and go" between the two wire lengths. However the westerly directivity of the longer wire continues to stand out. In Section four the performance of the longwire begins to deteriorate as compared with that of the short wire except for its due west pickup. Section five favors the short wire. Even the comparison westerly pickup levels off. However something was learned in switching between the two antennas for the 1390-kHz frequency. The longer antenna did produce the clearer signal. There was more "hash" in the short antenna pickup from other signals on the frequency. Thus the actual minus one reading is a debatable one indeed. The longwire deterioration in this segment indicates its length begins to approach a half wavelength and consequently there is a higher impedance at the receiver end which is not favorable for the transfer of maximum signal current to the receiver input.

The long distance performance of an antenna is better evaluated at night. However, most listeners understand that S-meter readings are very misleading during the nighttime hours unless the received signal is very strong. Co-channel and adjacent-channel signals as well as noise components contribute to the net S-meter readings. As a result, a true reading of the desired carrier level is difficult to obtain. Nevertheless

Table 8-2. Long-to-Short Wire Antenna Performance.

				Long to Short Wire S-Unit Difference
Section 1 540–740 kHz Segment				
N	WARM	590 kHz	Scranton, PA	+5
E	WNBC	660 kHz	New York, NY	+8
S	WIP	610 kHz	Philadelphia, PA	+8
W	WHP	580 kHz	Harrisburg, PA	+8
Section 2 750–950 Segment				
N	WGY	810 kHz	Schenectady, NY	+5
E	WTTM	920 kHz	Trenton, NJ	+2
S	WPEN	950 kHz	Philadelphia, PA	–1
W	WEEU	850 kHz	Reading, PA	+6
Section 3 960–1160 Segment				
N	WGPA	1100 kHz	Bethlehem, PA	–1
E	WINS	1010 kHz	New York, NY	+1
S	KYW	1060 kHz	Philadelphia, PA	–3
W	WHYL	960 kHz	Carlisle, PA	+3
Section 4 1170–1370 Segment				
N	WKAP	1320 kHz	Allentown, PA	–5
E	WBUD	1260 kHz	Trenton, NJ	–2
S	WHAT	1340 kHz	Philadelphia, PA	–6
W	WHUM	1240 kHz	Reading, PA	+3
Section 5 1380–1600 Segment				
N	WEST	1400 kHz	Easton, PA	–7
E	WBCB	1490 kHz	Levittown, PA	–4
S	WDAS	1480 kHz	Philadelphia, PA	–6
W	WLAN	1390 kHz	Lancaster, PA	–1

the listening quality of the received signals tell us much. The full story cannot be told in a few paragraphs. However the general results can be summarized.

In signal reception from the westerly direction the longer antenna almost always dominates. Even when its received signal is apparently weaker there is improved clarity because of the lower co-channel and adjacent channel pickup. For example, some Friday and Saturday nights the signal from WSM Nashville, Tennessee, was so strong I could switch over to a wide-band reception and enjoy the Grand Old Opry live on a high-fidelity speaker. Make note that WNBC New York is only 10-kHz higher in frequency than WSM. In other directions these two antennas ran true to form. Lower frequency signals did better on the longer antenna and vice versa. However, the advantage of having two antennas was proven any number of times. Especially during the sunset hours I could often identify two stations on the same frequency much more readily because one would dominate on one setting of the switch; the other on the second switch position.

Two antennas can add versatility to your medium-wave reception. Now what would happen if you directed one wire south and the other one west? Give it a try.

8-5. MW Loaded Long Longwire Antenna

The previous topic presented a comparison between two longwire antennas. One was cut to a quarter wavelength at the high-frequency end of the medium-wave band; the longer one lower in frequency near 800 kHz. When using a quick-change switch the significance of cutting such a wire an electrical quarter-wavelength for the portion of the band to be favored was demonstrated. Also, the longer wire was more directive towards the west than the shorter one, and westerly signals were received with less nighttime "hash" on them.

In the previous test (Fig. 8-5) the longer wire was not highly directive at the low-frequency end of the band. Directivity was improved by adding 60 ft to the longer wire. The new length of 360 ft corresponds to a quarter wavelength in the 650 kHz region. Improvement was modest at the low-frequency end of the band.

The next step was to try a resistive termination on the far end of the longer wire. Three 750-ohm resistors were paralleled and connected between antenna end and a 4-ft (1.22-m) ground rod driven into the ground.

The results with the far end terminated improved directivity modestly over the mid-frequency section of the band. Again listening results were better than figures suggest. At any rate the added length and termination was a definite improvement toward the west. In addition, the ability to switch between two antennas of different characteristics was a help in identifying two stations on the same frequency.

It must be stressed that S-unit comparisons often read true and positive but at other times they may only offer clues to what is going on because of the disguised influence of variables. For example another station on the same frequency, even a weak one, can spoil your results. This is an obvious problem because you can hear that station. Some adjacent channel spill over or a malfunctioning transmitter quite removed from the test frequency can ride up your S-meter and you are not particularly aware of it. More noise pickup on one antenna can fool you when checking low-level signals. At night when the band is heavy with strong signals, overload is possible, and you may just as well forget about many S-unit comparisons. However at night you can "listen," and by switching antennas the results become quite obvious. In fact it is the day measurements in combination with the night listening that tells the complete story.

8-5-1. Enter the Long Longwire Antenna

The next step is an additional jump in length to 585 ft which is just about as far as I can go at the moment with the present setup. The quarter wavelength frequency is lowered to near 400 kHz. Good directivity is now indicated over the entire band as indicated by the higher ratio of westerly pickup along with some declines in the north and south directions. The results suggest real directivity begins when the length of a long longwire is longer than any electrical quarter wavelength at the very low-frequency end of the band.

A very pronounced improvement took place when the long longwire was terminated with a resistor. The westward pickup jumped and almost every other direction fell off. Now you begin to see some Beverage characteristics. The Beverage is a low multiwavelength longwire with a high directivity off its far end. The height above ground is 10 ft (3.05 m) or less. In fact, the use of a termination resistor may be the way of introducing some Beverage-like performance when you are hampered in just how far you can stretch out your antenna. Termination photographs are shown in Figs. 6-6 and 6-7.

The antenna is supported about 10 ft above ground by PVC pipes and is reasonably straight except where I had to make some turns and bends to bring the wire into the radio room. Can you find some space to do some long, longwire DXing? Even insulated wire laid along the ground will do it for you. Search the hamfests and flea markets for some long hunks of wire.

Chapter 9

Confined Space and Indoor Antennas

ANTENNAS are "wavelength conscious." They intercept the most signal when their length and configuration are appropriate to the wavelength of the arriving signal. Furthermore to make full use of the intercepted signal there must be a matched transfer of that signal to the receiver by way of a transmission line or directly by one end of the antenna itself.

9-1. Introduction

Antennas have a need to be long for best signal interception on the long wave, medium wave, and a good portion of the shortwave spectra. Effectiveness becomes less when your antenna must be mounted indoors or in a much confined exterior setting. Indoor reception is the most trying because a signal may not always be present where you put the antenna. A good antenna placed where there is little signal to intercept will do worse than a poor antenna that you were lucky enough to position at a good signal spot.

Thanks to more sensitive and selective receivers plus good preamplifiers and other accessories most antenna installations perform surprisingly well. However, there are a number of ideas and arrangements that can be used to give a measure of improvement to the confined space and indoor antenna installation. A number of these ideas are presented in this chapter.

9-2. An Easy-Up Wideband Vertical Antenna for a Small Lot

Don't overlook the vertical antenna as a solution to your small lot limitations. You may be able to erect one just outside the radio room window or some other location very near the house or garage. When this is done only a short length of transmission line from the antenna to the receiver is needed. Verticals perform well and radio hams all over the world have been using them successfully for years. In fact, for reception purposes the receiver matching problem is not nearly as difficult as that encountered by hams when they must match their transmitter output to the antenna. The vertical has an omnidirectional horizontal sensitivity pattern and good low angle vertical pickup, the right combination for shortwave listening.

This is a two-topic description of a limited space vertical antenna installation for the frequency range between 1.5 and 20 MHz. The basic vertical is shown on the left of Fig. 9-1. To its right is a special loading coil that can be used to enhance tropical band operation. The loading coil assembly will be described in the next topic.

A solution to acceptable multiband reception is to choose a vertical antenna length that is practical and will provide good average performance over the range of bands of interest. Over a period of time of experimentation with SWB verticals my results indi-

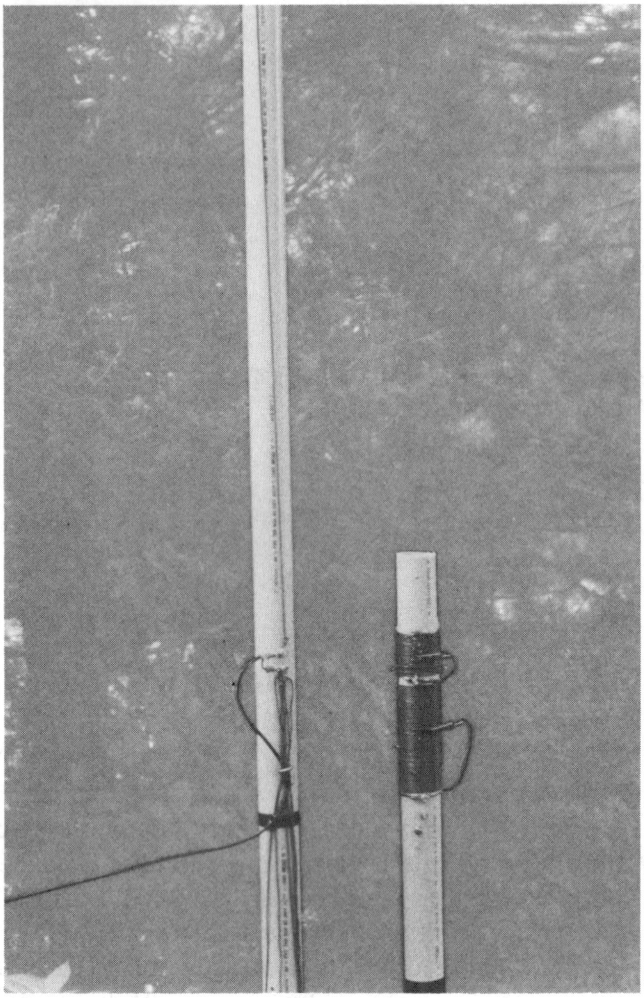

Fig. 9-1. Small-lot vertical and optional loading coil.

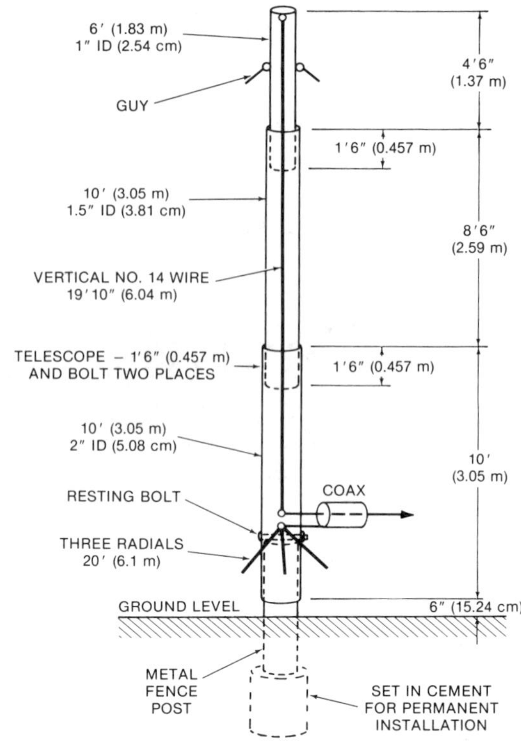

Fig. 9-2. Construction of a 25-meter vertical.

cate that a vertical cut to the 25 meter band is a valid compromise antenna. A quarter wavelength at this frequency corresponds to a length of 19 ft 10 in (6.04 m) (Fig. 9-2). The mast consists of three sections of PVC piping telescoped together as shown. This mast can be lifted and placed over a metal pole set into the ground. Mount this pole straight; if a permanent installation is desired, it can be set in cement. When the pole is set straight the plastic mast is self-supporting. If desired two or three rope guys can be used on the antenna to minimize whipping in the wind. The other ends of the guy ropes can be fastened to short stakes or other support points such as trees or buildings. Small stakes can be positioned about 10 ft (3.05 m) from the mast and occupy little space. Even less space is needed if you can bracket the mast to a building.

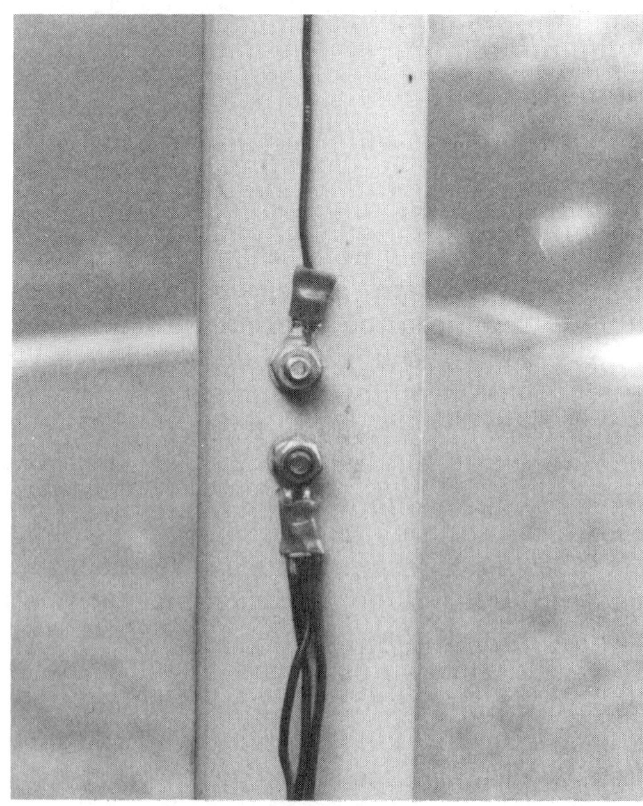

Fig. 9-3. Vertical antenna terminals.

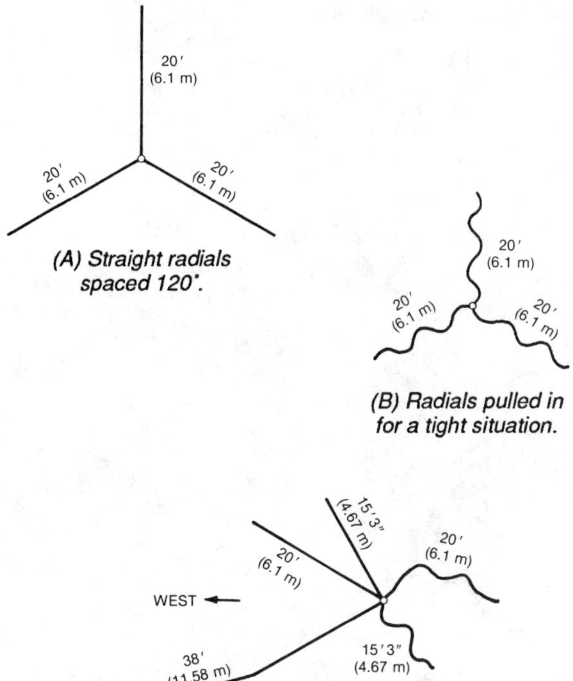

(A) Straight radials spaced 120°.

(B) Radials pulled in for a tight situation.

(C) Diagram of authors system.

Fig. 9-4. Radial system for vertical.

Three resonant radials complete the basic antenna, setting up a fine 25-meter "signal snatcher." As always, I choose to use No. 16 plastic-covered hookup wire for both vertical and radials. The wire is bared only where it connects to the two nut/bolt combinations that act as coaxial line terminals (Fig. 9-3).

Before the mast is erected the vertical wire is fastened to the top of the PVC mast. A solder ring is soldered to the bared opposite wire end and is held close to the mast by the bolt/nut terminal. The radials are run straight down to the bottom of the mast. At ground level they are fanned out and are buried just below the surface. If space is available, you can fan the radials out above ground starting right at the radial terminal; this improves results somewhat on the higher-frequency bands. Ideally, radials should be spread out with 120° separation whether above or below ground (Fig. 9-4A). However, things do not change too much if you must go out at differing angles. In fact, a serpentine arrangement of the radials can be used in a tight situation as shown in Fig. 9-4B. Perhaps you can run two straight radials and then a third one in an arc to accommodate lot arrangements and proximity to the house.

Additional radials can be used to extend the bandwidth of such a vertical by improving the match situation. In tests here good results were obtained by adding two additional 19-meter radials. Also, after some cut and try efforts, a single long radial was cut for 49 meters. By using appropriate curves and serpentines you can usually confine your radial radius to less than 15 ft (14.57 m) if that is necessary. Do the best you can, using just as much space as you have available.

My final arrangement is shown in Fig. 9-4C. Toward the house side I had to compromise the straight-out arrangement for one of the 19-and 25-meter radials. On the west side of the mounting position reasonably straight radials could be set down except that the long 49-meter radial was placed in a broad arc. A single 19-meter and 25-meter radial was placed above ground, winding it through nearby bushes about 3 ft (0.9 m) above ground. All the rest of the radials were buried.

In my situation, the signal sensitivity seems to be a bit better toward the west and southwest. It may be the proximity of the house and/or the cramped radials which cut back the northeasterly signals a small amount. Figure 9-5 shows all the various radials connected to the ground terminal as well as the transmission line. Finally silicon sealer was placed where the inner conductor of the coaxial line leaves the braid to prevent moisture from entering the line. All the exposed metallic surfaces except the outside nuts were covered with sealer to hold off early rusting and deterioration of the connections.

The vertical antenna installation in comparison with a number of reference antennas indicated good performance on all bands 19 through 49 meters. Results on these bands were within $\pm 1/2$ S-unit signal strength SINPO. (Recall that the S part of SINPO rating system refers to received signal level.) The results were fair on 13 and 60 meters and no less than -1 SINPO. On the remaining tropical bands the results were fair to poor. No tests were made on 11 meters. Thus, this antenna is quite a good performer for a tight space and simple antenna.

9-3. Loading Coil for the Small-Lot Vertical Antenna

The loading coils and matcher of Fig. 9-6 can boost the low-frequency performance of the simple multi-

Fig. 9-5. Vertical, radials, and transmission line connection.

Fig. 9-6. Vertical antenna and loading coils.

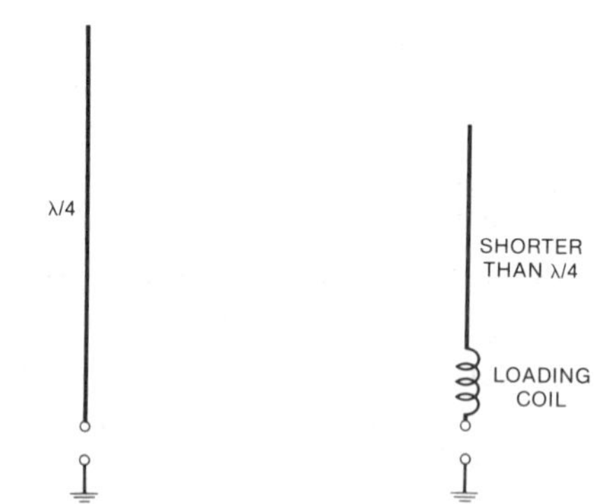

(A) Vertical antenna.

(B) Addition of loading coil to vertical antenna.

Fig. 9-7. Adding a loading coil permits use of a shorter antenna wire for a given frequency.

band small-lot vertical described in the last topic. You can also include MW/LW operation. A loading coil does not increase the signal pickup of a short vertical antenna; rather it improves the matching of the antenna to the transmission line. Whatever signal is picked up is then transferred from the antenna to the transmission line more efficiently. Thus, in the case of a quarter-wave vertical (Fig. 9-7) the physical length of the antenna can be shortened a reasonable amount and a loading coil inserted. Actually, at lower frequencies there can be a substantial shortening of the vertical and the signal delivered to the receiver can be little different and often better than that delivered by the usual random longwire used on the tropical bands.

A disadvantage of the loading coil is that it limits antenna bandwidth. Also, it is necessary to use coil taps that must be changed whenever you wish to peak the antenna to a given tropical band. For general listening though you can check for resonance on 75 or 90 meters and also do quite well on the other bands. You can also peak the system for the 30-, 40-, 80-, and 160-meter amateur bands, radio marine bands, and other UTE frequencies over the range between 1.5 and 9 MHz using the loading coil taps. The basic short vertical as described in Topic 9-2 provides good results from 9 to 20 MHz. In effect, with the addition of the loading coil you have an antenna that does very well from 1.5 to 20 MHz. Additional coils permit cov-

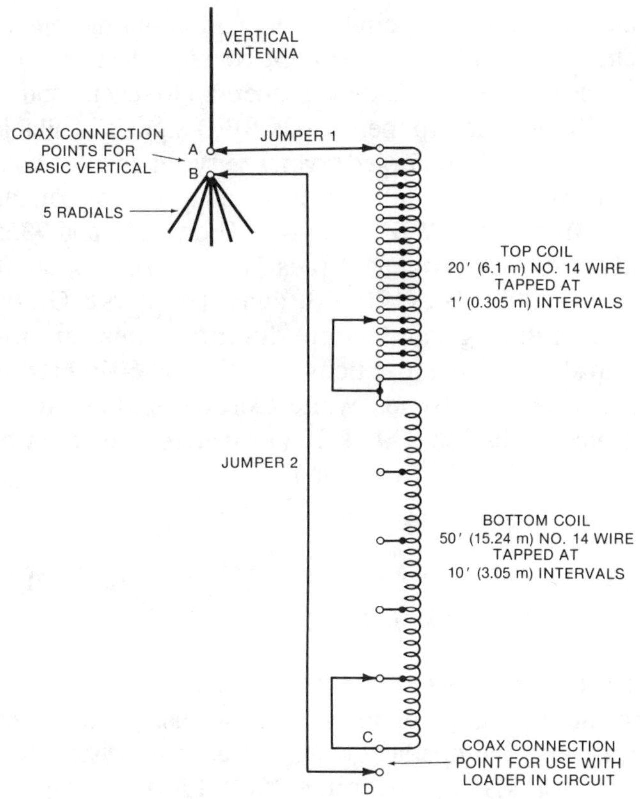

(A) Loading coil connections.

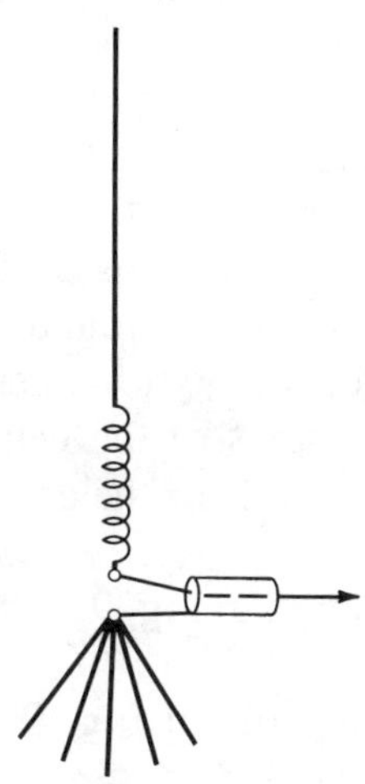

(B) Equivalent circuit.

Fig. 9-8. Connecting the loading coils.

erage from 100 kHz to 1.5 MHz. The loading coil wound on PVC piping in Fig. 9-6 was described in Chapter 4 (Topic 4-7).

When the loading coil is used, the terminals of the antenna system are located beneath the coil. In Fig. 9-6 you can see the transmission line connected to these terminals. The diagram of Fig. 9-8A shows how the loading coils are connected into the circuit. Jumpers 1 and 2 do the job. In fact, Jumper 2 can be left in the circuit permanently. Jumper 1 connects the top of the loading coil to the vertical antenna terminal A. Jumper 2 connects the radial system of the vertical antenna to terminal D at the bottom of the loading coil assembly.

When the loading coil is to be placed into operation, connect Jumper 1 into the circuit. Connect the inner conductor of the coaxial transmission line to terminal C and the braid to terminal D. In effect you have connected the loading coil in series with the vertical antenna and the ground radials as shown in Fig. 9-8B. A close look at Fig. 9-6 shows Jumper 1 at the top. The transmission line beneath it connects across the two terminals at the bottom of the loading coil and, at the very bottom, Jumper 2 provides a path to the radial terminal of the basic vertical.

The loading coil is taken out of the circuit by disconnecting Jumper 1. The coaxial line must now be connected across the two antenna terminals (A and B) on the basic vertical. This restores the operation of the antenna system to that covered in Topic 9-2. Also visible in Fig. 9-6 is the alligator clip and lead that permits the selection of a proper tap on the bottom coil. As shown the bottom coil has been shorted out. If you look closely you can also see the alligator clip and lead that is used to locate the proper tap on the top coil. It is these two taps that must be adjusted when you wish to obtain the very best performance on a given tropical band or other frequency segment between 1.5 and 9 MHz.

9-2-1. Loading Coil Checks and Operation

The most precise way of finding an optimum tap position for a given band is to use an antenna bridge such as the MFJ Enterprises MFJ-204B connected across the antenna terminals at the bottom of the loading coil assembly. All you need do is to start with the highest frequency band (in this case 31 meters) because less inductance is needed to resonate the 25-meter vertical on 31 meters. In my installation the antenna bridge indicates 31-meter resonance with bottom coil L2 shorted and the top coil L1 wire connected to front tap One (numbering beginning at the top of the coil). Only a small amount of inductance is needed.

The next step is to locate a tap that adds a bit more inductance to resonate on the 41-meter band.

Table 9-1. Sample of Loading Coil Tap Positions†

Band	Coil L1*	Coil L2
SWB Bands		
120 Meters	Open	Tap 2
90 Meters	Tap 8F	Tap 1
75 Meters	Tap 8F	Short
60 Meters	Tap 6F	Short
49 Meters	Tap 4R	Short
41 Meters	Tap 3R	Short
31 Meters	Tap 1F	Short
Ham Bands		
160 Meters	Tap 5F	Open
75 Meters	Tap 8F	Short
40 Meters	Tap 4R	Short
30 Meters	Short	Short
Marine Bands		
2.2 MHz	Open	Tap 3
4.2 MHz	Tap 7F	Short
6.3 MHz	Tap 4R	Short
8.4 MHz	Tap 2F	Short

*F = Front of Coil (L1), R = Rear of Coil (L1)
†Higher Frequency SWB, Ham, and Marine Bands best received with basic vertical and loading coils out of circuit (Jumper 1 open)

Table 9-1 shows that the right amount of added inductance is found on rear tap Three. Coil L2 remains shorted.

Table 9-1 also shows the various tap positions for all the SWB bands—31 through 120 meters—plus various ham and marine bands. Your settings, more than likely, will not be the same as mine but the table can give you a clue as to the vicinity of the best possible taps. You can do a reasonable job of locating the precise taps by listening if you set up some kind of a comparison antenna. It is important that you prepare such a chart to avoid a repeat of the necessary procedure for locating good tap positions when you change bands.

Table 9-1 is a sample showing tap positions that peaked my installation on the various bands. They may differ for your installation. You may wish to number your taps in sequence when you prepare the chart and label each tap on your coil. In my table, I numbered the taps from the top of the coil to the bottom facing the coil (F) as I approached the coil from the walkway. I did the same for the coil taps located on the opposite side (R) of the loader mast. Using this approach the table makes it easy to find the tap for each band without circling the coil counting taps or attaching numbers to them. Of course, I had to circle the coil initially to find the proper tap for each band.

In addition to the various SWB signals available for testing, don't forget several very reliable 24-hour time signals that fall within or near to most of the low-frequency SWB bands—CHU on 3330 and 7335 kHz (90 and 41 meters) plus WWV on 2500, 5000, and 10,000 kHz (120, 60, and 31 meters). On 49 meters during the daytime doldrums there are the Canadian broadcast stations on 6005 and 6070 kHz. A nighttime Deutsches Welle German signal can be heard reliably on 3995 kHz (75 meters) plus a lot of radio hams on the same band.

9-3-1. Adding Medium-Wave and Long-Wave Coils

The tapped coils of Fig. 9-6 provide antenna operation to the high edge of the AM broadcast band. When both coils are operated open, the antenna system resonates at some point in the 1500–1700 kHz region. Additional coils can be constructed to resonate the antenna system at the high-, middle-, and low-frequency portion of the broadcast band as well as the LW spectrum. An appropriate coil is connected across the two top terminals that link the bottom of the vertical and the top of coil L1 as shown in Fig. 9-9. These coils replace Jumper 1 as shown in Figs. 9-9 and 9-10.

Four coils were wound. On the left of Fig. 9-11 is the coil for the high frequency end of the broadcast band. Two copies of the center coil were wound. A single one is used for the center of the broadcast band and two in series for the low end of the band. The third coil shown in Fig. 9-11 was wound for LW operation.

The left coil in Fig. 9-11 is a winding of 20-gauge plastic-covered wire, random wound on a plastic coil form. Actually, the coil form is a wire spool with a center winding diameter of 1-in (2.54-cm) diameter. Radio Shack sells their hook-up wire on such spools.

The width of the slot is 5/16 in (0.794 cm) and the slot depth is 1/2 in (1.27 cm). Fill the slot with the random winding. The exact number of turns is not critical so long as you fill up the slot space with wire.

Two of the center coils were wound using the same type of form and the same filled-slot random

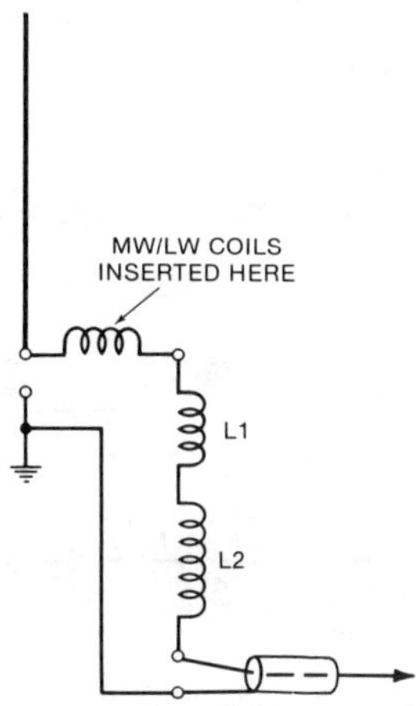

Fig. 9-9. Insertion spot for medium-wave and long-wave coils.

Fig. 9-10. Coil for the high-frequency end of broadcast band.

winding. However, the wire used is a very thin plastic-covered stranded type such as 24 gauge. I used some old Belden 8014 indoor antenna wire. Litz wire would be good if you can find some.

The coil form for the LW coil has the same dimensions except that the width of the slot is 2$^1/_8$ in (5.4 cm). Use the same type of wire random wound to fill the form.

The LW coil resonated the antenna system on approximately 350 kHz. You can move the resonant frequency up a limited amount with any coil combina-

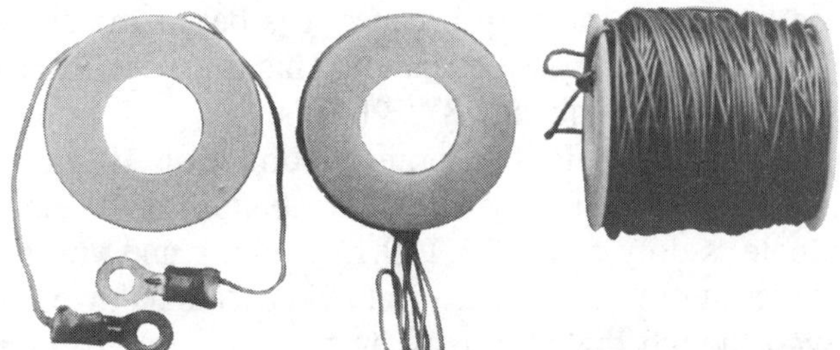

Fig. 9-11. Set of medium- and long-wave coils.

tion in the circuit by using one of the L2 tap positions instead of operating L2 completely open.

This confined space vertical with its matching coil and loader as well as extra loading coils can give you a good-performing system as well as a reference antenna from 100 kHz up to about 20 MHz.

9-4. Indoor Antennas

Variables and inconsistencies are the rule in the installation and test of a suitable indoor antenna for shortwave listening. Move the antenna a bit and the results can be quite different, especially on the higher frequencies of the shortwave bands. Antenna tests seem inconclusive and even resonant points for specific lengths of antenna wire become vague. Sometimes an antenna wire taped to the ceiling does no better than the same length of wire laid beneath the carpet. All this is related to the nearness of so many metallic surfaces, some of which are attached to the power system ground and others which are not. It is a hodgepodge of interaction.

Despite these limitations all is not chaos. Improvements can often be made and sometimes results are surprisingly close to the performance of a modest outdoor antenna.

Shielding of the signal can be a severe problem and is usually prevalent in high rise buildings because of their extensive steel girder support system. In a large city the very nearness of similar buildings adds to the problem. Usually an antenna system performs better if it can be positioned near a window to perk up signal pickup from at least one or two key directions. Even aluminum siding in city, suburb, or country sites can do a good job of shielding the signal from the indoor antenna. Sometimes a long antenna in the attic or crawl space does a fine job. Even a long antenna in the basement can help because it is below the siding and signals can pass beneath it. This is especially true on the lower frequency SWB bands.

Antenna height is more important on 19 meters and higher frequency shortwave bands. A 19-meter dipole is only about 30-ft (9.14-m) long and you can stretch it out in the attic or crawl space of a dwelling even though the ends may drape vertically. You can obtain fine results on this favorite DX band along with good results on 16 and 25 meters. Remember that a good DX antenna on the higher-frequency bands need not be too long.

Use No. 18 or No. 20 hook-up wire for your indoor antennas. The smaller diameter insulated wires are more flexible and can be hidden away. The coaxial feed line can be run along the baseboards where you bring it down from above or up from below. If space is available, you may wish to try a dipole on 25 or 31 meters. In an apartment try a thin wire dipole placed along the hall at baseboard level or in and out of two rooms. For a large room, the ends can be folded back along the side walls at right angles. Usually it is easier to disguise the antenna than the coaxial line that must be linked back to your receiver. However television has made us more accustomed to the coaxial lines that are snaked about our dwellings.

Indoor single wire antennas often produce surprising results although they are more erratic and more hit-or-miss experiments are necessary. In the case of a single-wire antenna the total quarter-wavelength measurement must be made from the antenna terminal of the receiver to the far end of the antenna wire. Again, use insulated hook-up wire for the antenna wire and you will avoid the problem of trying to avoid other metallic surfaces in the house that may come in contact with the wire.

A good ground is advisable from the standpoint of better signals and, especially, less noise and interference pickup. A good ground exists in the form of a heavy wire that runs to the nearest cold water pipe. However, in terms of signal pickup it has been a rare occasion when there was any real significant difference in S-meter readings when the ground is removed from the receiver.

There are three comparable indoor antennas in operation at my location. One is a 250-ft (76.2-m) crawl-space antenna that zigzags back and forth along the rafters for about 30 ft (9.14 m). It provides SW/MW/LW reception. A second single-wire antenna for SWB reception only is 48 ft (14.63 m) long (a quarter wavelength on 60 meters) and drops down under the carpet from the radio table, passing across the hall and into the living room where it stretches out under the carpet into a large L shape. It performs very well despite the fact that it is near ground level of the bungalow. A third antenna was a rather elaborate tunable (but very short) antenna in the radio room. This is my reference antenna for checking indoor antennas. Its construction and operation will be detailed at the end of this topic.

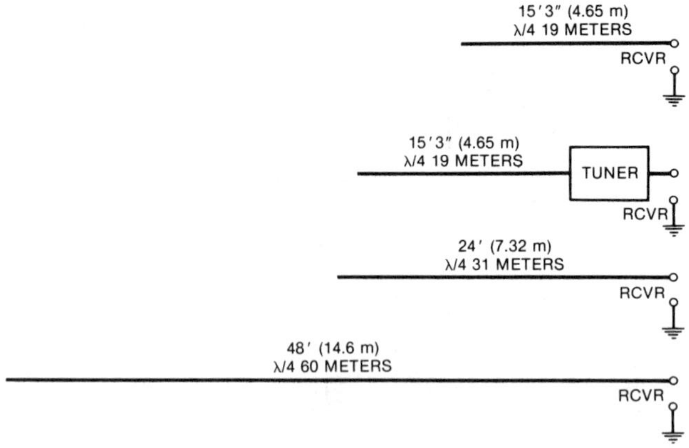

Fig. 9-12. Some basic indoor antenna lengths for single-wire antenna.

Quite often an indoor antenna consists of a 15–20-ft (4.57–6.1-m) length of wire (Fig. 9-12). Actually a length of 15 ft 4 in (4.66 m) corresponds to a quarter wavelength on 19 meters. Reasonable results should be obtained on this band. It should also do well on 25 meters. However you can do better because short wires of limited length are very much affected by surroundings; namely, the level of signal present at its physical location. Also the performance of such a short wire is poor on the lower-frequency bands. Many of the same problems exist in situations where the wire length is less than 25 ft (7.62 m). A quarter wavelength on 31 meters is 24 ft 1 in (7.35 m) and operating conditions remain mediocre. You must position the wire at just the right spot to obtain acceptable results. It's a matter of cut and try. In the right spot a 24-ft (7.32-m) length also does well on 41 and 49 meters.

It has been my experience that a tuner is of no use, or only of limited help in building up signal level from a short antenna on the higher frequency SWB bands. However, such a tuner does help in transferring more signal to the receiver input from the antenna on the lower-frequency bands despite the shortness of the antenna wire.

My own experience has indicated that an antenna length of about 48 ft (quarter wave on 60 meters) does very well and is a good compromise between excessive wire length for indoor use and poor performance. The longer wire length is of help in deriving good low-frequency band signals and, at the same time, a third harmonic (3/4 wavelength) relationship boosts the performance on the high-frequency bands, taking away some of the placement problems encountered with the very short antennas.

Stretch it out in as straight a line as possible. However, an L shape does well and even a U if you keep the two legs of the U separated as far as possible. Certainly this longer indoor antenna gives the outdoor 50-footer (15.24 m) at 18-ft (5.49-m) height a good chase. Also a tuner can be of definite help in boosting signal levels on the 75-, 90-, and 120-meter bands and on up to the medium wave band.

A large ungrounded metal mass such as a metal desk or floor to ceiling metal rack when used as an antenna can give a boost to signals on bands 31 through 120 meters. Higher-frequency bands did better on just the length of wire I ran to the rack. I put an alligator clip on the wire end so I could favor either end of the spectrum.

A "flexo" switch provides additional versatility for an antenna installation because it is able to compensate for placement variables. (Refer to Chapter 5.) An indoor installation using 33-ft and 48-ft antenna wires did well for me. However, you can choose any pair of indoor antenna lengths you like. Mount the switch close to the receiver and you can always make a quick changeover of the switch setting when you change bands. Inasmuch as most antennas display some directivity you will find that the direction of arrival of the signal on occasion influences the switch position that delivers the best signal to the receiver. Often the switch is useful in choosing a position with the cleanest signal when there is on-channel or adjacent channel interference.

9-4-1. Use of the Loading Coil

Often the length of an indoor antenna wire must be restricted severely. Such a short wire by itself responds poorly to the great range of frequencies allocated to shortwave broadcasting. Even on the higher-frequency bands, such as 19 meters, where its length can be cut to one-quarter wavelength the wire may not intercept much signal because of poor placement and shielding. The same short antenna outdoors would do well on high frequencies.

There is one technique to be considered as a means of adapting a short antenna for good all-band reception. A loading coil can be attached to the end of a short 10–12-ft length of wire. Such an arrangement tunes the antenna to resonance and helps to transfer as much signal as possible to a short length of line that connects to the receiver input. Additionally a series of coil taps is used to find an optimum coil turns for each individual shortwave broadcast band. The loading coil was described in Chapter 4.

Such an arrangement became our indoor reference antenna. The antenna assembly is positioned at the operating table and tap positions can be changed with ease while experimental indoor antennas are tested band by band. The loading coil is an excellent antenna by itself if you do not mind making the tap changes. Its all-band performance duplicates the 48-ft (14.63-m) wire antenna.

The tapped loading coil antenna is rather elaborate but it does result in good indoor antenna performance. For a shielded location, position the assembly and especially the antenna wire near a window. Perhaps you can permit a good portion of the wire to be on the outside of the window. When aluminum siding is the problem you may be able to find a way to attach some of the antenna wire to some exterior metallic surface.

9-5. Loop Antennas

Two basic wire-loop antennas are air-wound around a frame or wound around a ferrite core. In the first type (Fig. 9-13A) the turns of wire are wound around a wooden or plastic frame which can be square, rectangular, or wedge shaped. Typically loops are 2–4-ft (0.61–1.22-m) across. Larger loops would provide more pickup of desired signal but there is more difficulty nulling out strong locals and other undesired signals. The avid BCB DXer uses a loop to null out strong locals permitting reception of weak signals

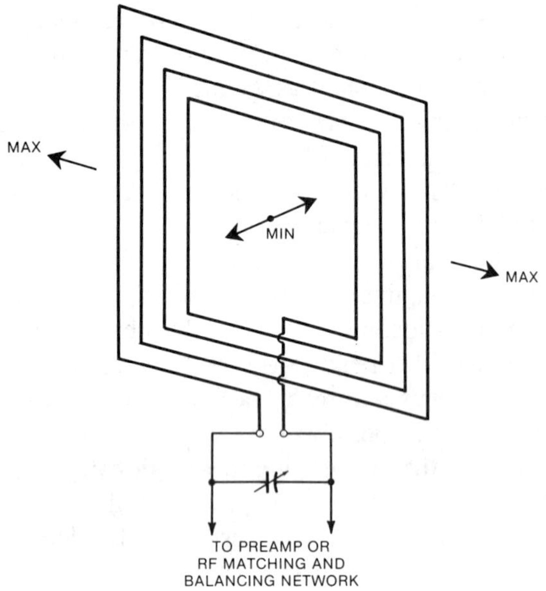

(A) Air-wound Loop

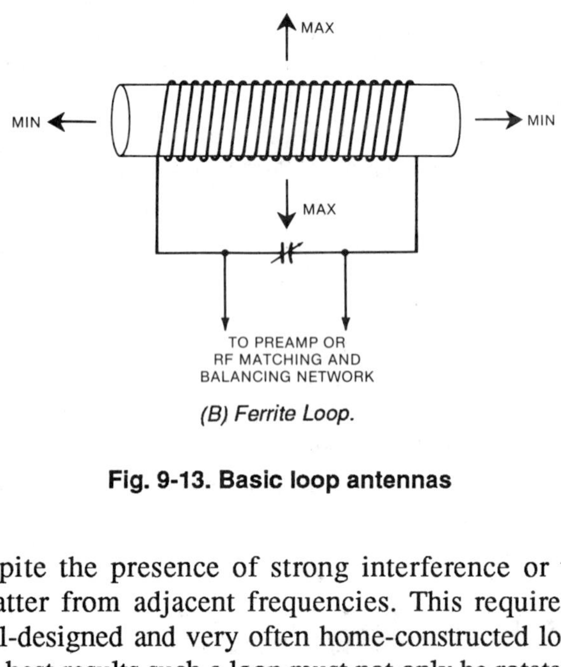

(B) Ferrite Loop.

Fig. 9-13. Basic loop antennas

despite the presence of strong interference or the splatter from adjacent frequencies. This requires a well-designed and very often home-constructed loop. For best results such a loop must not only be rotatable but must be able to be tilted vertically to obtain a deep signal null.

Another loop application permits the reception of two similar-level signals on the same frequency but with bearings that are divergent. Simply rotate the loop to null-down one signal to receive the second one and vice versa. Loop design is not so critical for this application except when the angles of signal arrival are near to each other or from exactly opposite directions.

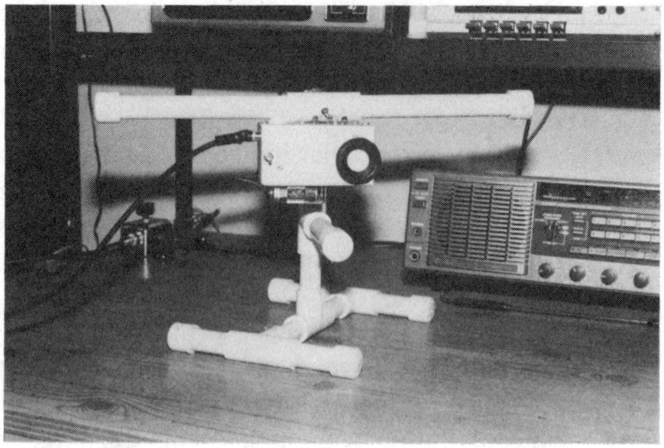

Fig. 9-14. Radio west ferrite loop and preamplifier.

An air-wound loop displays a null when its plane is perpendicular (broadside) to the direction of signal arrival. Maximum pickup is in line with the loop frame. The ability to tilt the loop helps to deepen the null.

The second loop type uses a coil wound around a ferrite rod (Fig. 9-13B). Tests and comparisons continue to be made and there are advocates of both loop types. The ferrite loop antenna displays a null along the line of the ferrite rod. Maximum pickup is broadside to the rod. Small, compact, and effective ferrite loops are practical.

A fine-performing, small loop construction is the "Radio West Little Loop," pictured in Fig. 9-14. The ferrite rod coil and its surrounding plastic cylinder extend along the top of the assembly. A balanced, high-gain preamplifier in a shielded case is attached to the plastic cover. Its output connects to the antenna input of the receiver. The entire assembly is rotated by the single plastic arm beneath the amplifier. The ferrite rod can also be tilted.

The loop performs well and deep nulls are obtainable. Of course all other antennas must be disconnected from the receiver when the loop is used. I did better by also disconnecting the receiver ground as well.

Basically this is a MW 540- to 1600-kHz loop. However my unit tuned between 400–2300 kHz and although gain fell off above and below these frequencies acceptable results were possible over an additional span of frequencies. The loop was pulling in WWV on 2500 kHz and the null was pronounced and accurate in the WWV direction. Reasonable results were obtained over the high frequency half of the LW band. The antenna did well on the 160-meter ham band, too.

All you need to do is peak tune the signal you wish to null with the tuning capacitor of the loop. Now rotate the loop to a signal null. When working toward the best possible null it often helps to use the most receiver attenuation that will still give you a readable signal from the DX station you are trying to identify in the null.

Chapter 10

VHF/UHF Receiving Antennas

IN the VHF/UHF spectrum the antennas are short and are usually constructed of tubing and rods. Wire antennas are scarce although they can be low cost and effective. The VHF allocations are between 30–300 MHz while UHF is generally considered to be spread between 300 and 1000 MHz although technically the UHF band extends from 300 to 3000 MHz. Frequencies above 1000 MHz are usually called microwave frequencies. Dimensions are usually given in inches and centimeters for the VHF/UHF spectrum. Key dimensions for the VHF-Lo, FM broadcast, Aviation, VHF-Hi, UHF-Lo and UHF-Hi are given in Chapter 11 (Table 11-3). Dimensions for those radio amateur bands in the same VHF/UHF spectrum are presented in Chapter 15.

10-1. Introduction

The five most common VHF/UHF antennas are depicted in Fig. 10-1. First is the quarter-wave ground plane which is the most common mobile antenna and a very popular listening post antenna. In mobile installation the car body serves as a ground plane. In other installations the ground plane is usually made up of a series of radials. These radials are resonant; that is, they are cut to resonance at the same frequency as the main vertical element. Notice from the tables in Chapter 11 that the length of a quarter-wave vertical decreases from 92.3 in (234.5 cm) at 30 MHz to 3.13 in (7.95 cm) at 885 MHz.

The dipoles shown in Fig. 10-1B are sometimes used. Most often the vertical height is a quarter-wave radiator and a quarter-wave "skirt" beneath it permits the dipole to be center fed with a coaxial line running up through the center of the skirt. In this arrangement no radials are needed. Ordinary vertical dipoles are used, too. Horizontal dipoles are used for the reception of the horizontally polarized component radiated from FM broadcast stations. At other times the horizontal or vertical dipole is the driven element of a parasitic beam as in Fig. 10-1C. Such an antenna is directional and must be oriented in a preferred direction. Refer to Chapter 6.

A beam usually has a single parasitic reflector and one or more parasitic directors. A multielement antenna of this type is called a Yagi antenna, Fig. 10-2. The gain and directivity increase with the number of directors. Yagi antennas can be mounted with vertical polarization as shown or with horizontal polarization.

The collinear antenna Fig. 10-1D is also a vertical. However, it is substantially longer than a quarter-wave or a half-wave type. A ground plane is not needed. Such an antenna has gain as a result of its low-angle vertical radiation pattern. The horizontal radiation pattern remains omnidirectional similar to the quarter-wave vertical.

The vertical collinear antenna uses in-line antenna elements positioned one above the other. Such an array is a very practical one for the VHF/UHF spec-

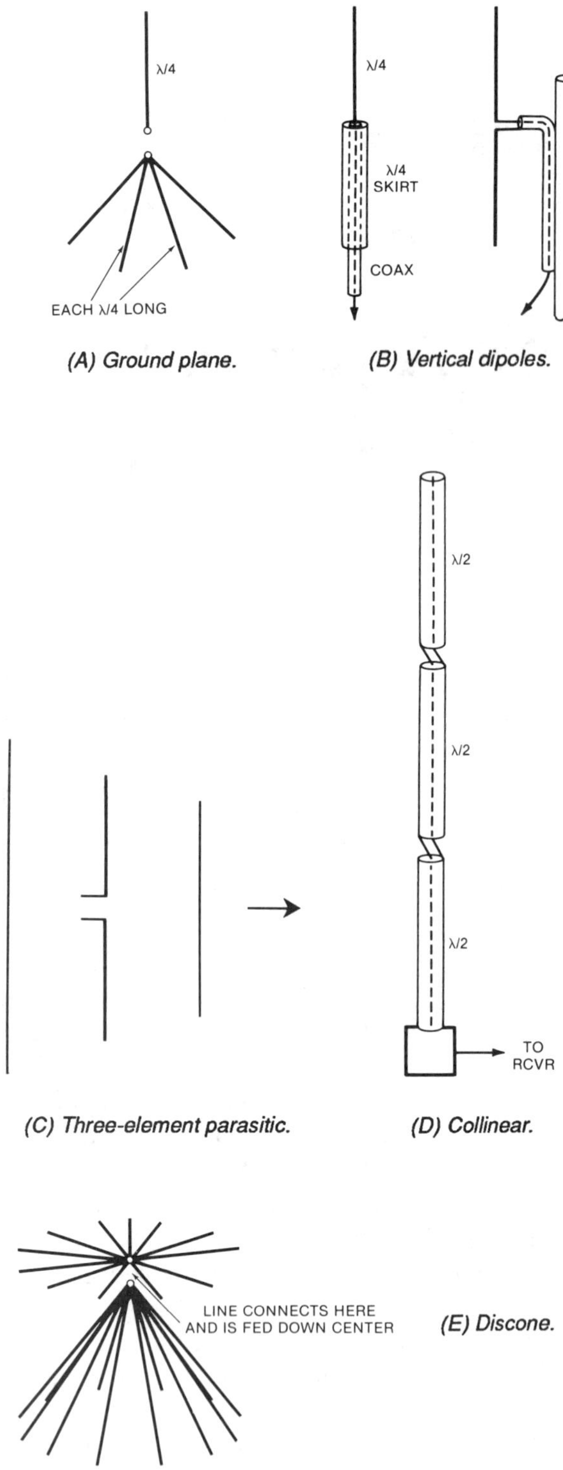

(A) Ground plane. *(B) Vertical dipoles.*

(C) Three-element parasitic. *(D) Collinear.*

(E) Discone.

Fig. 10-1. Basic VHF/UHF antennas.

Fig. 10-2. The UHF Yagi antenna.
(Courtesy Cushcraft Corp.)

Fig. 10-3. A VHF collinear antenna.
(Courtesy Celwave Systems, Inc.)

trum because of the short physical length of quarter-wave and half-wave elements at these frequencies. For proper operation such elements must be fed in phase. Figure 10-1D shows a simple method of doing so for a collinear comprised of three half-wave vertical elements. An impedance matching arrangement is needed at the base. Note that in the two separated areas between the coaxial verticals the feed is transposed—the inner conductor of the lower element connects to the outer conductor of the element above. Hence the half-wave elements are fed in phase as required. The entire assembly is enclosed in fiberglass and has the appearance of a simple rod antenna.

The VHF collinear antenna of Fig. 10-3 provides a 3-dB gain above a vertical dipole on the VHF-Hi band. Its length approaches 9 ft (2.74 m). Note from Table 11-3 that a quarter wavelength on this band is only about 2 ft (0.61 m).

Figure 10-1E is a wideband vertical antenna called a discone. It consists of a disc and a large cone that permits impedance matching over a wide span of frequencies. Discone antennas of practical size can be made to operate over a wide frequency span—for example, between 100 and 900 MHz. The disc and cone structure can be made of solid metal, tubing, rod, and even hardware cloth.

In the topics that follow a variety of antennas for use in the various VHF/UHF radio services are described. Some homebrew antennas are covered also. Remember that although dimensions given are for a particular band you can use the dimension tables to come up with a similar arrangement for some other band.

10-2. VHF/UHF Antenna Testing

The best way to test a scanner or any VHF/UHF antenna installation is by on-the-air-listening and a system that includes provisions to switch quickly between two antennas. One of the antennas can be an all-band omnidirectional scanner antenna mounted permanently at one spot. This antenna will be your reference comparison antenna. Presently I have such a reference mounted on a PVC pipe clamped to my roof vent pipe. Space the antenna or antennas to be compared at least 30 ft (9.14 m) away; mine is about 60 ft (18.3 m). Use approximately the same antenna height and length of transmission line; this is not an overly critical matter but just be reasonable.

The setup in the radio room is shown in Fig. 10-4. The antenna switch is a Radio Shack (15-1249) push-button VCR/TV/FM switch for 75-ohm lines. This switch uses female F-connector inputs and outputs, permitting a complete 75-ohm line and F-connector installation.

Unless you want to modify your scanner and add an S-meter circuit you must test results aurally. Aural

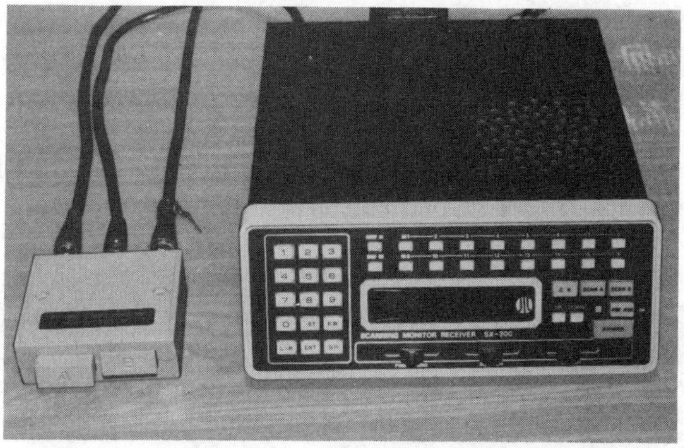

Fig. 10-4. Scanner and coaxial antenna switch.

techniques work out very well but you must know what to listen for. You must also be patient and wait for stations in test directions to come on the air and, then, make quick changeovers to evaluate performance.

Testing on strong high-level signals is the most difficult. When signals are strong they produce complete receiver quieting even though the signal levels from the two antennas may differ significantly. This indicates that you should not test on locals or the very strong more distant stations. However you can sometimes get a glimmer of what's happening by setting your squelch full on. This is the point that requires a strong signal to break through. You will hear both signals at the same level. But pay no attention to the signal. Instead listen for background noise level or rush. You may hear that the background-level quieting is not as good with the signal from one of the antennas as it is from the other. This indicates that the signal from one antenna is below the level of the other.

For a weaker signal you may wish to set the squelch near the point at which noise begins to open the squelch circuit. At this point you can make a comparison between two signals picked up on the antenna. The stronger signal may break the squelch while the weaker signal will be broken up and may not reproduce without distortion.

When there is a significant difference in signal levels the difference in antenna performance is apparent immediately. For example a weak signal coming from a distance might easily break through the squelch and reproduce well. However the same weak signal picked up on the reference antenna may not be heard at any setting of the squelch control. It is this difference in performance on the weaker signals that shows clearly how well one antenna performs in comparison to another. Keep a frequency record of these signals so you can go back to them time and time again. In one check on 450 MHz I could hear a distant signal well on one antenna and nothing at all when I switched to a second antenna. In another test I could read a signal from the Philadelphia airport tower, albeit weakly, while the same signal picked up from a second antenna broke up. These two examples showed a decided difference in the pickup of the two antennas. When two antennas do not differ greatly it is more difficult to give the preference to one antenna over the other. Slight as it might be such a difference, on occasion, may be a readability factor in the reception of a very weak signal.

10-3. Simple Mobile Antennas

On-the-road scanner listening has become a popular facet of the hobby. It is easy to contrive a simple, low-cost, and versatile antenna. Start with a luggage rack or gutter mount antenna. At a sacrifice in versatility you can go with an FM mobile antenna and its permanent cowl, fender, or roof mount. Some of these telescope, especially those for combined FM/VHF/UHF application. Their length can be adjusted to match a quarter wavelength on a particular band. However, they are physically short in terms of peak reception on the 30–50 MHz band.

Dimensions for one-quarter and three-quarter wavelength vertical antennas for the landmobile bands are given in Chapter 11 (Table 11-3). This table also gives the dimensions for some special segments such as aviation and marine allocations.

Many enthusiasts use the same scanner in the radio room and in the auto. Often the scanner sits on the front seat or dash and power is derived with an appropriate plug and cable that can be inserted into a dashboard cigarette lighter. A short length of transmission line links the antenna to the scanner by way of the popular F-connector. An older station wagon is my radio car because various antenna mounts can be attached to its luggage rack. All sorts of antenna components can be carried including mast sections. My preference for on-the-road listening is when the car is parked (more enjoyable and safer) so I can go with a more elaborate or higher antenna when useful. The site is often one of higher elevation, too. If you plan to listen mobile, why not choose a favorable location?

Fig. 10-5. Antenna mounted on front crossbar of luggage rack.

A helpful antenna has been a telescoping AM/FM unit designed as a gutter mount for a truck (Radio Shack 12-1324). It fits the crossbar of my luggage rack and can be positioned right above the rear door as shown in Fig. 10-5. The short transmission line has an easy entrance to the vehicle and can then be draped over the front seat. Look your vehicle over and make certain you will have a point of attachment for the antenna. If not, look over other types of mounts to find the one most suitable. Antenna length extends to almost 30 in (76.2 cm) and its telescoping feature permits you to adjust it to a variety of suitable lengths for the various bands. It is a bit short for the 30–50 MHz band but when extended fully provides some pickup. It does very well on the marine and aviation bands. Furthermore if you adjust the height to 17–18 in it performs well on the 150- and 450-MHz bands. On the higher-frequency band the antenna functions as a three-quarter wavelength vertical with good results. Another convenient antenna is the Grove Enterprises ANT-10 (Fig. 10-6) which has a magnetic base mount that will hold firmly on any metallic auto surface. It operates well on 150- and 450-MHz bands. The helical winding helps on 50 MHz.

Fig. 10-6. Scanner vertical antenna with magnetic base.

Fig. 10-7. Luggage rack mount for 30–50-MHz vertical.

A second versatile antenna mount is one designed for luggage rack mounting (Radio Shack 21-937). Basically it is a CB antenna mount designed for whips or other CB antenna types with a 3/8-in thread that will screw into the mount. It can be seen mounted on the rear cross bar of a luggage rack in Fig. 10-7. A good scanner antenna can be obtained by chopping off the whip to obtain a total length of 69 in (175 cm). Note from Table 11-3 that the length corresponds to a quarter wavelength on 40 MHz. If your interest is aviation radio this antenna also performs well as a three-quarter wavelength vertical on the band. Pick-up is good on other frequencies also.

If you like, you can shorten the antenna to about 54 in (137 cm) and still have acceptable performance on the 30–50 MHz bands. This shorter dimension provides a three-quarter wavelength cut at the center of the landmobile and marine bands. Remember that its length is such that it can reach up into some tree branches when you drive! The best bet is to stop at a good spot, screw it in, and listen.

Another possibility if you have no particular interest in 30–50-MHz listening in the auto is to form your own vertical from a rod, stiff wire, or tubing. The example in Fig. 10-8 shows a stiff rod shaped with a small loop at one end. A 3/8-in large-head screw was purchased at the hardware store which along with a lock washer completed a good mobile antenna for marine band scanning and general 150-MHz band reception. The length of the antenna was 18 in (45.7 cm) which also provided good results on the 450-MHz band as well.

10-4. The VHF/UHF Discone

The discone antenna is a wideband low-impedance antenna. In the early designs by A. G. Kandoian (Fig. 10-9) there was a large sheet-metal base cone which served as the ground plane, and above it, a sheet-metal disc. The inner conductor of the coaxial line connects to the center of the disc and the braid connects to the cone. The Kandoian design dimensions in

Fig. 10-8. A 150-MHz vertical antenna constructed from a stiff rod.

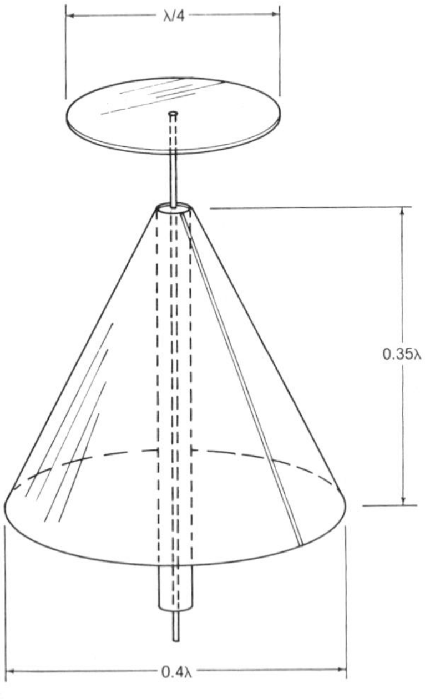

Fig. 10-9. The basic solid disc-and-cone (discone) antenna design.

Fig. 10-10. The Encomm Diamond D-130 discone antenna mounted on roof.

Fig. 10-9 are based on the lowest frequency of operation. Proper design can sustain a constant antenna impedance up to a frequency ten or more times higher than the lowest frequency of operation.

Similar results are possible with the use of rod or tubing. The Encomm Diamond D-130 antenna (Fig. 10-10) is an example of a design that operates from 25 to 1300 MHz. The discone section consists of eight 11-in rods that form the disc plus eight 33-in ones that comprise the cone. The base separation is about 33 in. To extend operation down to 25 MHz there is a vertical element plus a base loading coil assembly that screws into the top of the basic discone. The discone assembly arrangement is depicted in Fig. 10-11. Also note the base loading coil and its vertical rod element.

Fig. 10-11. Discone assembly arrangement.

Fig. 10-12. The Antenna Specialist Discan antenna.

The disc and cone serve as a ground plane for low-frequency operation.

The antenna performs well on all the landmobile and associated bands. It's great to have one antenna that works well on so many bands over such a wide frequency span.

The Antenna Specialist Discan™ in Fig. 10-12 is a low-cost version that uses the same basic principle. It operates over a frequency range of 25 and 512 MHz with fine results on the aviation, VHF-Hi, UHF-Lo and T-bands (T-bands are segments of the UHF-TV band assigned to the landmobile radio service). VHF-Lo band reception is down in spots probably because of dips in the omnidirectional pattern as compared with a full-fledged discone. There are three disc rods 20 ft 6 in long that branch off at 120° from the center plate and mounting bracket. The cone rods are 44 ft 6 in long. A coaxial mount is positioned at the center of the plate and U-bolt bracket.

10-5. T-Band Scanner Yagi

In a number of metropolitan areas across the country the landmobile radio services have been allocated frequencies on unused UHF TV channels. Channels are selected which will not cause interference with the area UHF TV stations. The so-called T-Band section extends between 470–512 MHz and includes TV channels 14 through 20. It is best to call your local FCC district office to obtain the specific assignments for your region.

Scanner listening on these assignments can be improved by using one of the regularly available, low-cost commercial UHF-TV antennas, particularly the corner reflector and Yagi types. Mount them to obtain vertical polarization instead of horizontal. This will usually require some reworking of the mounting hardware. The results will surprise you especially if you live in the near or far fringes of a metropolitan area. I live in the near fringe of Philadelphia and by putting a Yagi into operation I receive signals well that are not there at all when I switch over to a multiband omnidirectional scanner antenna.

An antenna I converted very simply was a Winegard 7-element 75-ohm UHF Yagi with a gain of

(A) U-bolt saddle.

(B) Antenna mounted on pipe with revised U-bolt.

Fig. 10-13. Adapting the U-bolt assembly for operation on vertical polarization.

Fig. 10-14. Top-of-mast arrangement and transmission line.

Fig. 10-15. Corner reflector mounted with vertical polarization.

12 dB. The model was obtained from R.F. Electronics (refer to addresses in Appendix). These Yagis have an approximate bandwidth that covers 10 adjacent UHF-TV channels or more than 60 MHz. Hence one Yagi chosen for operation on channel 16, 17, or 18 provides coverage on the entire landmobile T-band. Purchase one that best matches the one or two allocations of your metropolitan area. Find out what channels are active in your area. Assignments are not only made in the city proper but include the surrounding countries as well. Such a Yagi mounted in the central area of the city will do well for omnidirectional pick-up. In addition you can mount it to display maximum sensitivity toward a scanner-active suburban area. The ultimate would be the use of two or more switchable Yagis or even a small TV rotor. You could easily mount a small 3-element 150-MHz Yagi on the same mast as well.

The first step is to change the Yagi over to vertical polarization. On the Winegard array this can be done easily by rotating the U-bolt saddle 90°. First, drill out the support rivet which can be seen at the center of Fig. 10-13A to free the saddle. You can now remove and rotate the saddle. Drill the new holes required to fasten the saddle and provide U-bolt insertion.

Figure 10-13B shows the Yagi mounted on a

Fig. 10-16. Modified UHF-TV corner reflector.

short 3-ft section of 1-in PVC piping. This Yagi uses a split-element Gamma matching section with a female F-connector permanently attached. Fasten a male F-connector to the antenna end of your transmission line. Figure 10-14 shows the transmission line attached with a weather shield pulled over the F-connection. The short 3-ft (0.91 m) section of mast is telescoped into the top of the main mast.

10-6. UHF Corner Reflector Antenna

The corner reflector (Fig. 10-15) is a fine UHF directional antenna with good gain and wide bandwidth. One cut for the UHF-TV band covers the landmobile, T-band, and more, doing well even down in the 430–450-MHz ham band.

For landmobile operation the UHF-TV antenna must be changed over to vertical polarization. In my test we selected the inexpensive Radio Shack 15-1629 antenna. In changing over to vertical polarization, the U-bolt assembly was abandoned. However the mounting plate of the corner reflector has two convenient holes that permit a vertical mount. All you need do is to drill corresponding holes in the metal or PVC mast and use two bolt/nut/lockwasher pairs. The mounting arrangement is shown in Fig. 10-16. Since the antenna impedance of the corner reflector is 300 ohms you can use a Radio Shack 15-1143 (4-to-1) matching transformer, also visible in Fig. 10-16, for a match to a low-impedance coaxial line.

Fig. 10-17. Two-element marine band beam.

In operation there is a significant gain over my reference discone in the direction of maximum pick up of the corner reflector. Of course the corner reflector results were poorer than the reference antenna on the lower-frequency landmobile bands.

10-6-1. Modifying Antenna for Marine Band

Some of the ardent scanner listeners like to change antennas rather often. What to do with the discarded antenna? Well, come up with some kind of a gain or special antenna that you would find useful. After taking my good Grove Enterprises OMNI antenna out of operation I made it into a two-element beam for operation in the marine segment of the VHF-Hi band (Fig. 10-17). The antenna is cut for about 159 MHz operation. Figure 10-18 shows the dimensions. You could also cut it for operation in the aviation band. Refer to the dimension tables if you wish to cut an antenna for some other band or frequency.

The two driven elements are cut back to the appropriate lengths for marine band operation. The same OMNI feed arrangement is kept intact. You can use any dipole support insulator you may have available. A reflector is cut and positioned 0.2 wavelength

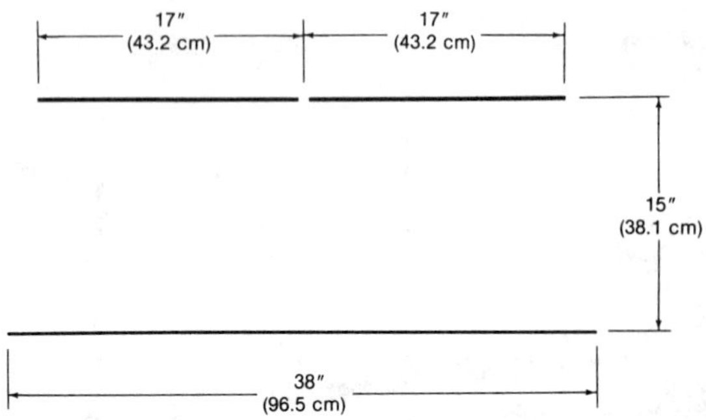

Fig. 10-18. Two-element beam dimensions.

Fig. 10-19. Addition of reflector to the beam.

behind the dipole (Fig. 10-19). The spacing chosen keeps the antenna feedpoint impedance from dropping too low for attachment to 50–72-ohm lines. An appropriate reflector mounting hole is drilled through the crossarms and the mounting is stiffened with two back plates from a pair of old U-bolt assemblies found in the junk box. Keep a look-out for old antenna parts in junk shops, trash collection pick-up points, flea markets, and hamfests. You can often find enough materials to build up a variety of antennas with the aid of the dimension tables in Chapter 11.

Fig. 10-20. Multielement FM beam with three driven elements.

This small beam provides a modest gain above my reference discone in its maximum direction.

10-7. FM Broadcast Antennas

Multidriven element and basic Yagi antennas provide sharp unidirectional patterns and high gain for long-distance reception on the FM broadcast band. Extra dollars must be invested in a rotator and a sturdy mast or tower. On occasion a rotator may not be needed if your objective is to beam on some far away metro area, or just in some one favored DX direction.

The antenna of Fig. 10-20 consists of a reflector, two directors, and three driven elements. The use of two or more driven elements minimizes the wideband matching problem. The antenna impedance is about 300 ohms instead of the low value of a regular Yagi which requires a matching network that can restrict bandwidth. You can use 300-ohm twin lead with the antenna or feed a coaxial line by using a 4-to-1 matching transformer.

Often a simpler and less costly solution is attractive. There are two common problems. One possibility is that you live in a small city or town a modest distance from a metro area and you would like a good gain and a broad pickup pattern toward the target. A second possibility is that you live in a metro area and you would like some cut back in the received signal levels from the metro locals as you aim your antenna toward more distant stations.

In such situations you may wish to use the small two-element minibeam of Fig. 10-21. It has some gain, good back rejection, and will operate well with-

(A) FM minibeam.

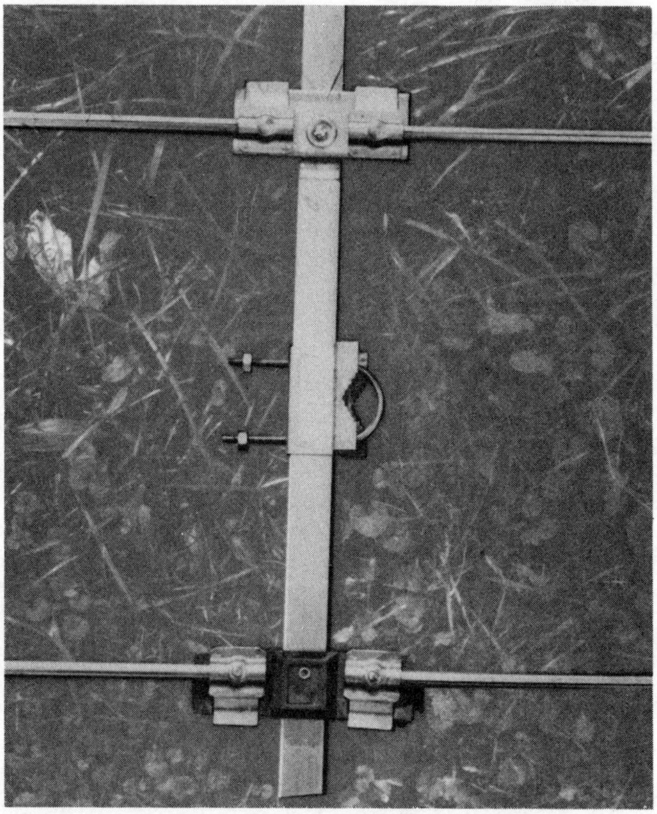

(B) Assembly.

Fig. 10-21. FM minibeam and simple construction.

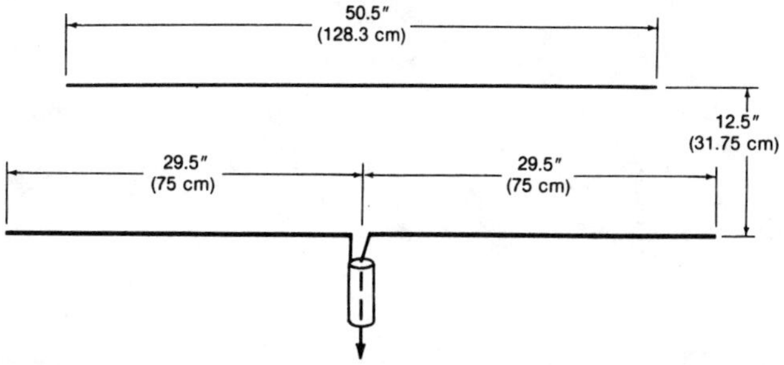

Fig. 10-22. FM broadcast minibeam dimensions.

out a matching device. A dipole and parasitic director combination were chosen because the front-to-back ratio is somewhat better than for a dipole and reflector pair. The director spacing is about 0.1 λ and, although the antenna impedance drops low, the back rejection is good and mismatch to a 50-ohm line is not too serious. It is a compact antenna and requires little space. Figure 10-22 gives the dimensions.

One example of its performance is a clean changeover between WBYO (Boyertown PA) and WBLS (New York City) with 180° rotation (easy arm-power rotation of a PVC mast). Both operate on 107.5 MHz. WBYO is about 20 miles due west; WBLS is about 70 miles ENE. When the beam is anchored ENE, the New York and New England reception is fine and there is minimum interference from Philadelphia stations on adjacent channels. For example, I can receive a clean WQXR signal on 96.3 MHz despite WWDB on 96.5 MHz. My FM receiver is a GE Superadio II. It is a superb AM portable but not especially great on FM. Thus, it makes a good receiver for putting FM antennas through tests.

If you have been reading a number of chapters of this book you know my preference for more than one antenna for active DXing. FM listening is not an exception. Usually there are three antennas active here. A three-position coaxial switch does the job. Interesting changeovers occur, especially during DX openings. A recent session included the minibeam, the Diamond discone and the small-lot vertical of Topic 9-2. Remember it is always the strongest signal that takes over and that could very much depend on the antenna in use at the moment. Good DXing all.

Chapter 11

International Antenna Dimension Tables for Radio Listeners

THE dimensions listed in Tables 11-1, 11-2, and 11-3 permit you to look up practical lengths for antenna elements, spacings, and matching sections as an aid in the construction of antennas for use on the various shortwave broadcast, scanner, and CB bands. Table 11-4 gives a set of practical constants and equations that you can use to calculate the same information for any specific frequency not given in the tables. This could be for some spot frequency within any one of the bands or any frequency over the entire MW, HF, shortwave, or VHF/UHF spectra. The first two columns in Table 11-4 give the constants for calculating dimensions in feet and meters. These columns are normally used for the MF, HF, and shortwave bands. The last two columns give the constants for calculating dimensions in inches and centimeters. Thus, these columns are normally used for VHF and UHF applications. In the MW and HF listing (Tables 11-1 and 11-2), dimensions are given in feet and meters. For the VHF/UHF spectrum (Table 11-3) dimensions are in inches and centimeters.

11-1. Dimension Tables

The first two columns of Table 11-1 show the frequency in megahertz and the wavelength in meters. Next is the column for the length of a quarter-wave vertical or one segment of a dipole. The top figure is in feet and inches and the bottom figure is in meters. The fourth column presents the length of the radials to be used with a quarter-wave vertical antenna.

The fifth column supplies the dimension for each 3/4 wavelength segment of a 3/2 wavelength longwire antenna. The reflector and director columns give the lengths of parasitic elements that are a part of two- or three-element beams. For a three-element beam, the driven element must be longer than that of a dipole. This required length is found in column eight.

Next are three free-space dimensions useful in spacing the parasitic and driven elements of a directional antenna. The final two columns list the lengths of quarter-wave matching sections for coaxial lines with velocity factors of 0.66 and 0.79. The 0.66 figure is for standard coaxial lines; the 0.79 figure is appropriate for foam-dielectric coaxial lines and close enough for some flat lines such as 300 ribbon.

Tables 11-2 and 11-3 give the same information for their respective bands, except the frequency is listed in kilohertz for Table 11-2.

Some examples of table usage follow:

1. What element lengths are needed for the construction of a 60 meter dipole? Refer to the meters column and the 60-meter row in Table 11-1. The answer is 48 ft. Two such lengths are needed for a dipole.
2. An antenna for operation on the aviation band requires a coaxial quarter-wave stub for matching. What is the required stub length if foam-dielectric coaxial line is to be used. Refer to the

Table 11-1. Antenna Dimensions for International Shortwave Broadcast, Scanner, and CB Bands*

Freq. (MHz)	Meters	λ/4 Element	λ/4 Radial	3/4λ Element	Reflector & Free Space λ	Director	Beam Driven Element	Free Space λ	Free Space 0.25 λ	Free Space 0.2 λ	λ/4 Stub 0.66 VF	λ/4 Stub 0.79 VF
2.4	120	98' 29.71	100' 30.48	303' 92.2	205' 62.5	185' 56.7	100' 30.6	410' 125	102'6" 31.25	82' 25	67'8" 20.6	81' 24.7
3.3	90	71' 21.61	73' 22.17	220' 67.06	149' 45.5	135' 41.2	73' 22.3	298' 90.9	74'6" 22.73	59'8" 18.18	49'2" 15.00	58'11" 17.95
3.95	75	59' 18.05	61' 18.52	184' 56.03	125' 38.0	113' 34.4	61' 18.6	249' 75.9	62'3" 18.99	49'10" 15.19	41'1" 12.53	49'2" 14.99
4.9	60	48' 14.55	49' 14.93	148' 45.16	100' 30.6	91' 27.8	49' 15.0	201' 61.2	50'2" 15.31	40'2" 12.25	33'2" 10.10	39'8" 12.09
6	49	39' 11.88	40' 12.19	121' 36.88	82' 25	74' 22.7	40' 12.3	164' 50	41' 12.5	32'10" 10	27' 8.25	32'5" 9.87
7.2	41	32'6" 9.9	33'4" 10.16	101' 30.74	68'4" 20.8	61'10" 18.89	33'6" 10.21	137 41.67	34'2" 10.42	27'4" 8.33	22'7" 6.874	27' 8.23
9.7	31	24'1" 7.35	24'9" 7.54	75' 22.81	50'9" 15.46	45'11" 14.05	24'10" 7.58	101'5" 30.9	25'4" 7.73	20'4" 6.186	16'9" 5.102	20' 6.11
11.8	25	19'10" 6.04	20'4" 6.2	61'6" 18.75	41'8" 12.71	37'9" 11.53	20'5" 6.23	83'5" 25.4	20'10" 6.356	16'8" 5.085	13'9" 4.194	16'6" 5.02
13.7	21	17' 5.20	17'6" 5.34	53 16.15	35'11" 10.95	32'6" 9.93	17'7" 5.37	71'10" 21.9	18' 5.474	14'4" 4.38	11'10" 3.612	14'2" 4.323
15.3	19	15'4" 4.66	15'8" 4.78	47'5" 14.46	32'2" 9.80	29' 8.89	15'9" 4.80	64'4" 19.61	16'1" 4.9	12'10" 3.922	10'7" 3.235	12'8" 3.871
17.7	16	13'3" 4.03	13'7" 4.133	41' 12.5	27'10" 8.475	25'2" 7.684	13'7" 4.15	55'7" 16.95	13'11" 4.237	11'1" 3.39	9'2" 2.80	11' 3.346
21.6	13	10'10" 3.301	11'1" 3.387	33'7" 10.245	22'9" 6.944	20'7" 6.296	11'2" 3.40	45'7" 13.89	11'5" 3.472	9'1" 2.778	7'6" 2.291	9' 2.742
25.8	11	9'11" 2.764	9'4" 2.835	28'2" 8.578	19'1" 5.814	17'3" 5.271	9'4" 2.849	38'2" 11.63	9'6" 2.907	7'8" 2.326	6'4" 1.918	7'6" 2.296
27.1 (CB)	11	8'8" 2.63	8'10" 2.70	26'10" 8.166	18'2" 5.535	16'5" 5.02	8'11" 2.71	36'4" 11.07	9'1" 2.768	7'3" 2.214	6' 1.826	7'2" 2.186

*Dimensions in feet and meters

125 MHz row in Table 11-3. The last column states 18.66 inches. Conversion Table 11-5 indicates that 18.66 inches is very near to 18 11/16 inches.

3. What is the length of an end-fed quarter-wave longwire on the MW frequency of 1200 kHz? Refer to the 1200-kHz row of Table 11-2. The longwire should be 195 ft in length.

11-2. Practical Constants and Equations

Radio and light waves travel approximately 300,000,000 meters per second. If this quantity is divided by frequency you will obtain the wavelength of a radio wave in meters:

$$\lambda = \frac{300{,}000{,}000}{f}$$

where

λ is the wavelength in meters,
f is the frequency in kilohertz.

A more practical equation is one that can handle frequency in megahertz or:

$$\lambda_{Meters} = \frac{300}{f_{MHz}}$$

Because one meter is equivalent to 3.28 ft, the wavelength in feet would be:

$$\lambda_{Feet} = \frac{300 \times 3.28}{f_{MHz}} = \frac{984}{f_{MHz}}$$

Refer to the Free Space λ column in Table 11-1 and to the constants of Table 11-4.

The preceding equations are the basis for all the constants given in Table 11-4. To obtain the length in feet, inches, meters, or centimeters, as the case may be, of any antenna, element, or spacing simply divide the associated constant by the desired frequency in megahertz.

For example what is the practical length in inches of a reflector for operation on 153 MHz? The reflector constant from Table 11-4 is 5820. Find it. Therefore:

$$\text{Length} = \frac{5820}{153} = 38.04 \text{ inches}$$

Check it out in the 153-MHz row of Table 11-3.

What is the length in meters of one segment of a dipole antenna for operation on 7.2 MHz? The constant (from Table 11-4) is 71.3. Therefore:

$$\text{Length} = \frac{71.3}{7.2} = 9.9 \text{ meters}$$

Check it out in the 7.2-MHz row of Table 11-1.

What is the length in feet of a director for operation on 10 MHz? The constant (from Table 11-4) is 445. Therefore:

$$\text{Length} = \frac{445}{10} = 44.5 \text{ feet}$$

Table 11-6 gives a handy practical conversion from decimal feet to inches. Referring to it, you can determine .5 foot equals 6 inches. Thus, in our example 44.5 feet equals 44 feet, 6 inches.

Now you have calculated a length based on a frequency not listed in the dimensions tables. Easy! Is it not?

Table 11-2. Antenna Dimensions for Medium Wave Broadcast Band*

Freq. (MHz)	Meters	λ/4 Element	λ/4 Radial	3/4λ Element	Reflector & Free Space λ	Director	Beam Driven Element	Free Space λ	Free Space 0.25 λ	Free Space 0.2 λ	λ/4 Stub 0.66 VF	λ/4 Stub 0.79 VF
540	555	433	444	1344	911	824	446	1822	456	364	301	360
		132	135	410	278	251	136	556	139	111	91.6	109.7
640	469	366	375	1134	769	695	377	1538	384	308	254	304
		111	114	346	234	213	115	469	117	93.8	77.3	92.5
740	405	316	324	981	665	601	326	1330	332	266	219	263
		96.4	98.9	299	203	184	99.3	405	101	81.1	66.9	80.0
800	375	293	300	908	615	556	301	1230	308	246	203	243
		89.1	91.4	277	188	170	91.9	375	93.8	75	61.9	74.0
860	349	272	279	844	572	517	280	1144	286	229	189	226
		82.9	85.1	257	174	158	85.4	349	87.2	69.8	57.5	68.9
920	326	254	261	789	535	484	262	1070	267	214	177	211
		77.5	79.5	241	163	147	79.9	326	81.5	65.2	53.8	64.4
1000	300	234	240	726	492	445	241	984	246	197	162	194
		71.3	73.2	221	150	136	73.5	300	75	60	49.5	59.2
1100	273	213	218	660	447	405	219	895	224	179	148	177
		64.8	66.5	201	136	123	66.8	273	68.2	54.5	45.0	53.8
1200	250	195	200	605	410	371	201	820	205	164	135'3"	162
		59.4	61	184	125	113	61.3	250	62. 5	50.0	41.2	49.4
1300	231	180	185	558	378	342	185	757	189	151	125	149
		54.8	56.3	170	115	105	56.5	231	57.7	46.2	38.1	45.6
1450	207	161	166	501	339	307	166	679	170	136	112	134
		49.2	50.4	153	103	93.8	50.7	207	51.7	41.4	34.1	40.9
1600	187.5	146	150	454	308	278	151	615	154	123	102	121
		44.6	45.7	138	93.8	85	45.9	188	46.9	37.5	30.9	37.0
1700	176	137'8"	141	427	289	262	142	579	145	116	95'6"	114
		42.0	43.0	130	88.2	80	43.2	176	44.1	35.3	29.1	34.8

*Dimensions in feet and meters

Table 11-3. Antenna Dimensions for VHF/UHF Bands

Freq. (MHz)	Meters	λ/4 Element	λ/4 Radial	¾λ Element	Reflector & Free Space λ	Director	Beam Driven Element	Free Space λ	Free Space 0.25 λ	Free Space 0.2 λ	λ/4 Stub 0.66 VF	λ/4 Stub 0.79 VF
VHF Lo												
30	10	92.3	96	289.2	194.0	177.3	95.1	393.6	98.4	78.73	64.93	77.73
		234.5	244	734.6	492.8	450.4	241.6	1000	250	200	165	197.5
40	7.5	69.25	72	216.9	145.5	133.0	71.33	295.2	73.8	59.05	48.7	58.3
		175.9	182.9	550.9	369.6	337.8	181.2	750	187.5	150	123.7	148.1
45	6⅔	61.56	64	192.8	129.3	118.2	63.4	262.4	65.6	52.49	43.29	51.82
		156.4	162.6	489.7	328.5	300.3	161.0	666.7	166.7	133.3	110.0	131.6
FM BCB												
93	3.23	29.78	30.97	93.29	62.58	57.20	30.68	127.0	31.74	25.40	20.95	25.08
		75.66	78.66	237.0	159.0	145.3	77.93	322.6	80.65	64.52	53.21	63.69
98	3.06	28.27	29.39	88.53	59.39	54.29	29.11	120.5	30.12	24.10	19.88	23.80
		71.8	74.64	224.9	150.8	137.9	73.95	306.1	76.53	61.22	50.50	60.44
103	2.91	26.89	27.96	84.23	56.50	51.65	27.70	114.6	28.66	22.93	18.91	22.64
		68.31	71.02	214.0	143.5	131.2	70.36	291.3	72.82	58.25	48.05	57.51
Aviation												
125	2.4	22.16	23.04	69.41	46.56	42.56	22.82	94.46	23.62	18.89	15.58	18.66
		56.29	58.52	176.3	118.3	108.1	57.98	240.0	60	48	39.6	47.4
VHF Hi												
153	1.961	18.1	18.82	56.71	38.04	34.77	18.65	77.18	19.29	15.44	12.73	15.25
		45.99	47.81	144.0	96.62	88.32	47.37	196.1	49.02	39.22	32.34	38.73
159 M	1.887	17.42	18.11	54.57	36.60	33.46	17.94	74.26	18.57	14.86	12.25	14.67
		44.25	46.01	138.6	92.98	85.00	45.58	188.7	47.17	37.74	31.12	37.26
167	1.780	16.59	17.25	51.95	34.85	31.86	17.08	70.71	17.68	14.14	11.66	13.96
		42.13	43.80	132.0	88.52	80.92	43.4	179.6	44.91	35.93	29.63	35.48
UHF Lo												
460	0.652	6.022	6.26	18.86	12.65	11.57	6.202	25.67	6.417	5.135	4.235	5.07
		15.30	15.90	47.91	32.14	29.38	15.75	65.22	16.30	13.04	10.76	12.88
470	0.638	5.894	6.128	18.46	12.38	11.32	6.070	25.12	6.281	5.026	4.145	4.962
		14.97	15.56	46.89	31.45	28.75	15.42	63.83	15.96	12.77	10.53	12.60
490 T	0.612	5.653	5.878	17.71	11.88	10.86	5.822	24.10	6.024	4.820	3.976	4.759
		14.36	14.93	44.97	30.17	27.58	14.79	61.22	15.31	12.25	10.10	12.10
UHF Hi												
837	0.358	3.309	3.441	10.37	6.953	6.356	3.409	14.11	3.527	2.822	2.327	2.786
		8.406	8.74	26.33	17.66	16.15	8.658	35.84	8.961	7.168	5.912	7.070
850	0.353	3.259	3.388	10.21	6.847	62.59	33.56	13.89	3.473	2.779	2.292	2.744
		8.278	8.606	25.93	17.39	15.90	8.526	35.29	8.824	7.059	5.822	6.969
885	0.339	3.130	3.254	9.803	6.588	6.011	3.224	13.34	3.336	2.669	2.201	2.635
		7.950	8.266	24.90	16.70	15.27	8.189	33.90	8.475	6.780	5.592	6.693

*Dimensions in inches and centimeters

Table 11-4. Constants for Antenna Dimension Calculations

Dimension	Feet	Meters	Inches	Centimeters
Free Space λ	984	300	11808	30000
Free Space λ/2	492	150	5904	15000
Free Space λ/4	246	75	2952	7500
Free Space 0.2 λ	196.8	60	2362	6000
Driven Element λ/4	234	71.3	2770	7036
Reflector	492	150	5820	14783
Director	445	136	5320	13513
Beam Driven Element	241	73.5	2853	7247
3/4 λ Element	726	221.3	8676	22037
λ/4 matching section (0.79 VF)	194.3	59.23	2332	5923.5
λ/4 matching section (0.66 VF)	162.4	49.49	1948	4948.7
λ/4 Radial	240	73.15	2880	7315

$$\text{Length of Longwire Antenna (in meters)} = \frac{150(N - 0.05)}{f_{MHz}}$$

$$\text{Length of Longwire Antenna (in feet)} = \frac{492(N - 0.05)}{f_{MHz}}$$

where

N = Numbers of half wavelengths of the antenna

Table 11-5. Practical Conversion of Decimal Inches to Ruler Calibrations

Ruler Calibration	Decimal Inches
1/16	0.0625
1/8	0.125
3/16	0.1875
1/4	0.25
5/16	0.3125
3/8	0.375
7/16	0.4375
1/2	0.5
9/16	0.5625
5/8	0.625
11/16	0.6875
3/4	0.75
13/16	0.8125
7/8	0.875
15/16	0.9375

Table 11-6. Practical Conversion from Decimal Feet to Inches

Decimal Feet Range	Inch(s)
.01 – .03	0
.04 – .12	1
.13 – .21	2
.22 – .28	3
.29 – .37	4
.38 – .45	5
.46 – .53	6
.54 – .62	7
.63 – .70	8
.71 – .78	9
.79 – .87	10
.88 – .95	11
.96 – 1.00	12

Part 2

Ham Radio Antennas

Chapter 12

Antenna Fundamentals and Tests

IF you ever kept a record of topics of extended conversation on the ham bands you most likely would find antennas at the top of the list. More than any other technical activity antenna homebrewing and installation continue unabated.

12-1. Introduction

Part 2 of this book is for the ham antenna experimenter and installer. You need not be wealthy or have a super intellect to enjoy antenna experimentation. It is a pleasant endeavor and provides some outdoor exercise and, more often than not, the erection of the simplest antenna can be quite instructive and rewarding. In my own amateur activity, with its emphasis over decades on low height and low-cost antennas in association with low-powered transmitters, it has been possible to contact more than 100 countries more than once, and it has also been possible to maintain with satisfaction some worldwide schedules over the years.

12-1-1. Safety Again

Home construction and the installation of the antennas covered in this book are not hazardous if you obey very simple safety and property rules. These have been covered in the discussions of Chapters 1 and 2. Please review these chapters. Briefly, stay clear of power lines—they can be deadly. Be careful in lifting masts, stringing wire, and climbing. Think before you do and be alert when you act. Most of the antennas covered in this book can be erected from ground level. Of course, many of them, if you prefer, can also be supported on high masts and towers. Always seek help if you have had no experience with high installation procedures. Even if you have experience don't let up your vigil. Safety isn't automatic.

The topics that follow review some of the basic antenna types and relate them to ham operation and the use of the ham antenna tables of Chapter 16. Test and comparative procedures are discussed and some practical installations are described.

12-2. Basic Ham Antennas

The basic ham band antennas are the quarter-wave vertical, dipole, half wavelength sloper, and inverted dipole. These are the very same antennas that were described in Chapter 2 for radio listeners. Read and review the individual discussions.

In the erection and test of a ham antenna important details and procedures are involved. The transmit mode of operation is critical. There must be careful transmitter-to-feedline and feedline-to-antenna matching because a mismatch decreases the efficiency of the RF power transfer between the transmitter and the antenna. A serious mismatch can damage transmitter parts.

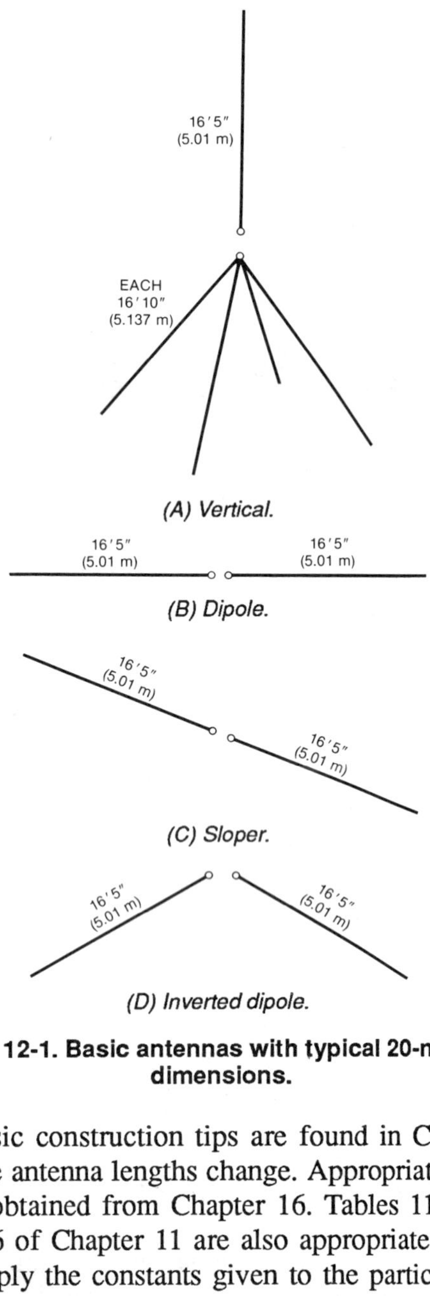

Fig. 12-1. Basic antennas with typical 20-meter dimensions.

Basic construction tips are found in Chapter 2. Only the antenna lengths change. Appropriate lengths can be obtained from Chapter 16. Tables 11-4, 11-5, and 11-6 of Chapter 11 are also appropriate. All you do is apply the constants given to the particular ham band frequency.

Several 20-meter antenna examples are given in Fig. 12-1. Note that the appropriate dimensions can be found in Table 16-1 under 20 meters. An appropriate design frequency is 14.240 MHz for sideband service. Proper values are found in columns 3 and 4. Compare the antenna types with Figs. 2-9, 2-18, 2-20, and Fig. 2-21. Except for the dimensions they are the same. The construction can be identical.

How do I cut one of the basic antennas for a specific frequency in the 20-meter band? The vertical and the inverted dipole will be used as examples.

To derive maximum benefit from the procedures it would be helpful to assemble a 20-meter vertical antenna initially. Except for dimensions it can be identical to the vertical of Figs. 2-9 through 2-11. If you mount your vertical at some appropriate location you may wish to keep it there permanently to use as a reference for checking all sorts of antenna types on the 20-meter band.

An antenna bridge and an SWR meter will be a help in checking out proper matching to the transmitter. The antenna bridge itself can be used to bring the antenna resonance rather near to a precise frequency. Then an SWR meter is brought into the picture to obtain a close match on some precise frequency such as 14.240 MHz.

12-3. Antenna Tune-Up

The tune-up procedure that follows is related to the 20-meter vertical of Fig. 12-1A and the inverted dipole of Fig. 12-1D. A factor of initial concern is to bring the antenna reasonably close to a 50–70 ohm match at the transmitter. My test recommendation is to use an antenna bridge for all preliminary tests. You can take your time and work systematically and you need not put a signal on the air. Actually most of the performance comparisons can also be made on reception. The next step is to tune up the transmitter at low power, matching the antenna system as closely as possible. Most modern transmitters permit you to do this initial tuning on a tune-up mode, which sets the output on low power in comparison with a maximum 200–250 watt rating. The antennas and procedures in this book are based on a power output of approximately 250 watts or less. Of course most of the basic antenna types covered can be operated at a higher power level with appropriate concern for antenna wire size, transmission line capacity, and other hardware ratings.

The popular standing-wave ratio (SWR) meter is calibrated to read the standing wave voltage on a transmission line. The usual SWR meter is based on transmitter load and transmission line impedance of 50 ohms and it indicates how well an antenna system matches this impedance. When there are no standing waves on the coax line (corresponding to a ratio of 1-to-1) the most efficient transfer of power exists between the transmitter and the antenna. As the SWR

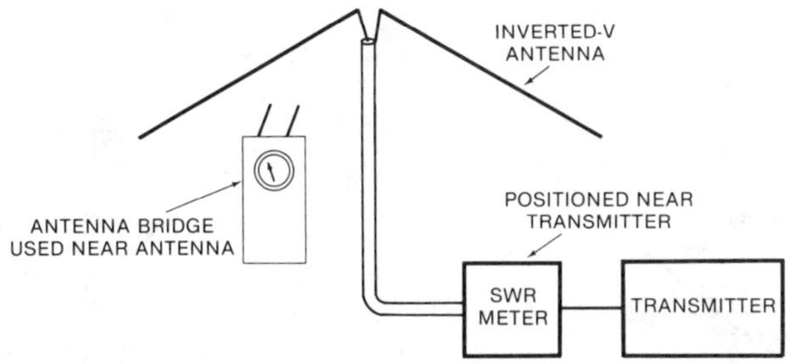

Fig. 12-2. Setup for antenna and transmitter tuning.

ratio becomes higher the efficiency declines. Ratios higher than 2-to-1 are considered too high and represent poor operating conditions. A 2-to-1 ratio causes about 10% of the RF power to be lost as a result of the line mismatch. There are other losses possible also as a result of the influence of such a mismatch on the operating conditions of the transmitter. Loss becomes considerably worse as the ratio rises above 2 to 1.

A problem in making power and SWR measurements in the past has been the influence of the instrumentation itself on the line operating conditions. The development of the insertion-type wattmeter designed for a specific impedance (usually 50 ohms) that can be left in the line permanently (Fig. 12-2) minimizes this problem. This meter reads direct power and reflected power and can be calibrated to read SWR as well. Often the instrument has both power and SWR display meters.

Such a meter designed for 50-ohm impedance operation is placed in series with the coax line between the antenna and the transmitter and has a minimum effect on the operation of the antenna system. Usually the SWR meter is mounted near the transmitter where it is visible for transmitter/antenna tune-up and monitoring services.

The vertical radiator is cut to the table value of 16 ft 5 in (5.01 m) and its resonant radials to 16 ft 10 in (5.137 m). The antenna bridge is attached at the antenna terminals of the vertical and indicates that resonance is likely to be on the low-frequency side of 14,240 kHz. The antenna bridge produced similar results when connected to the transmitter end of the transmission line, indicating an acceptable match of the line to the antenna.

The antenna bridge is used to bring the antenna to resonance somewhere in the band and also confirms a reasonable match when connected to the transmitter. This is a great help because the antenna tuning can be done with no on-the-air signals. Such preliminary tuning also avoids a serious transmitter mismatch.

A closer look at the match and resonance conditions can now be obtained with the SWR meter and transmitter. Make all initial SWR measurements at low power. Many transmitters provide SWR measurement on the tune-up power position. Follow the instructions provided with your transmitter and SWR meter. In this example the SWR reading was very near 1 to 1 at 14,150 kHz and was no more than 1.5 to 1 over the entire 20-meter band. Resonance was very near 14,240 kHz when the radiator length was cut back by 1.5 in (3.81 cm).

One factor should be kept in mind when trimming antennas. When the antenna is too long, the minimum SWR frequency occurs below the desired frequency. Trim a short length from the antenna and check again. When the antenna is too short, the minimum SWR occurs above the desired frequency. The antenna must now be lengthened. It is a wise procedure to initially cut the antenna slightly longer than the indicated table value. It is much easier to trim than to add wire to an antenna end.

The mounting arrangement of Fig. 12-3 suggests a way of fine tuning a vertical for resonance to two or more frequencies. In Fig. 12-3A the coaxial line connects between the top and middle antenna terminals. The radiator is cut for resonance near the high-frequency end (phone portion) of the 20-meter band. To operate at the low end (CW portion), connect a jumper of proper length between the top terminal and the middle terminal. Now the coaxial line is connected between the middle and the lower terminal as in Fig. 12-3B. Of course the radials must also be transferred from the middle terminal to the lower terminal. In my installation the terminals were accessible from a 5-ft (1.524-m) stepladder.

12-3-1. Inverted Dipole Antennas

One of the advantages of the inverted-V dipole is the ease with which the antenna ends can be dropped to near ground level for adjustment to some preferred resonant frequency. Thus, this arrangement allows a short length of wire to be clipped to each antenna end for shifting resonance from the phone to the CW portion of a given band. The same technique can also be used conveniently on any of the ham bands that are lower in frequency than the 20-meter band.

(A) Connections for high end of band.

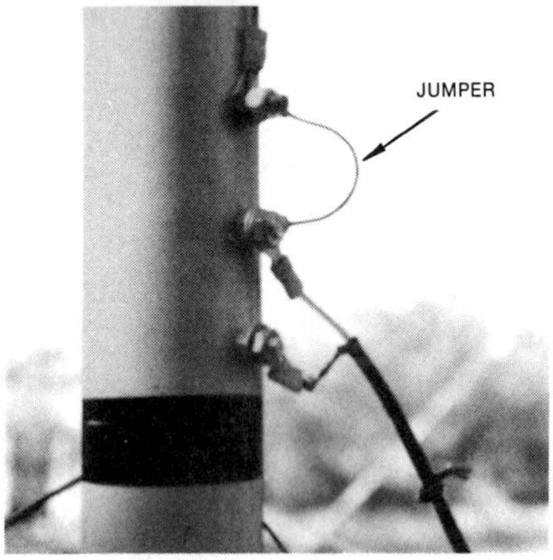

(B) Connections for low end of band.

Fig. 12-3. Jumper plan that permits operation at either end of the band.

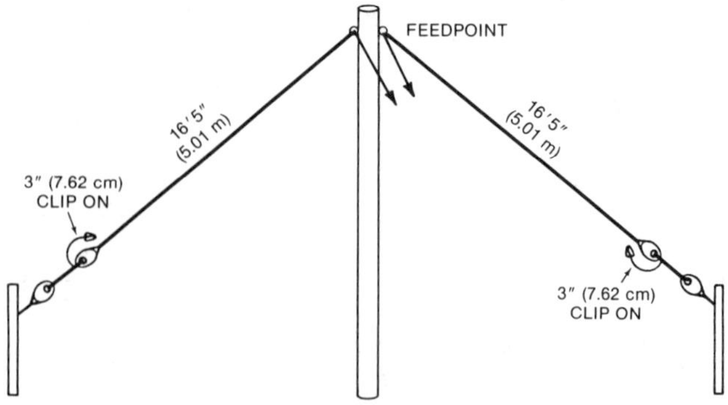

Fig. 12-4. Leg lengths and clip-on section that permit minimum SWR operation at low end of band.

Fig. 12-5. PVC mast top and antenna terminals.

My inverted dipole is shown in Fig. 12-4. The leg lengths are cut for 14,240 kHz resonance while the clip-on section can be used to lower this frequency to the DX end of the CW portion.

The bolt/nut antenna terminals are mounted at the very top of the mast as shown in Fig. 12-5. The two antenna wires and the coaxial line join here. Don't forget weatherproofing. Usually I connect a 20-ft (6.1-m) length of coaxial line. This arrangement makes it convenient to make resonance and impedance measurements of the antenna from ground level. A connector assembly is inserted at the ground end of the coax. This connector permits attachment to the longer portion of the coaxial line that links the antenna system to the radio room. Again, antenna bridge measurements are verified in the radio room before the line is connected to the transmitter output.

12-4. Testing Against a Reference Antenna

Ham antennas can be compared very well on receive-only mode provided proper matching exists. Both the antenna system under test and the reference antenna system must be matched to the antenna input of the transceiver or the transmitter/receiver pair. How systems compare on reception is also a reasonable indication of how they compare on transmission. I have often found receive checks to be more exacting when comparing antennas than collecting transmitter reports from stations that pick up your signal. For one thing, on reception you can make almost instant checks by switching quickly between two antennas several times and repeat the test at various times of day and for various distances and directions. A lot of propagation variables, such as differences in S-meter readings among the stations you ask for reports, human error, interference, and communications difficulties when conditions are poor, are avoided by this technique. In summary, careful *receive-only* tests provide a good indicator of just how well things will work out when you apply the transmitter output to the antenna. Of course the benefits of receive-only testing require that you have a reference antenna.

Transmitter tests using the antenna under test and the reference antenna can be used to verify your reception results. It is fun to do this. The comparisons may surprise you. Also, you will learn something about the tentativeness of received signal reports. The operator at the other end is trying to give you a truthful report. However, this operator is up against interference and propagation changes during the time you are going through the process of making measurements and keeping in contact so you can follow the necessary antenna changeovers in a smooth, rapid manner. Of course if the difference is quite decided between two antennas the problem is less severe. Doing it the receive-only way you can seek out the isolated signal and make your comparison instantly.

Various reference antennas can be used. Verticals can be easily assembled for bands between 30 meters and 10 meters. Inverted dipoles with segmented leg lengths do well for testing 20-, 40-, and 80 meters, and even up to 160 meters if antenna space is available.

In addition to the 20-meter reference vertical described previously, various multiband antennas with reference capability can be found in Chapters 14 and 15. The discone antenna covered in Chapter 10 serves well as a reference antenna for the VHF/UHF ham bands.

The comparison between the vertical antenna and the inverted dipole may be tentative. In a very general way the vertical is omnidirectional while the inverted dipole has maximum pickup and greater radiated power broadside to its length. Thus the inverted-V does better than the vertical in a broadside direction and poorer in-line with the wire.

In addition to their differing horizontal directivities there are several other important differences between horizontal and vertical antennas. One major one is that their comparative vertical angles of pickup and propagation are different. As a result there can be performance differentials that are not only influenced by the compass bearing between stations but also the station distance, along with changing propagation conditions which shift the vertical angle of arrival. On occasion the vertical antenna dominates the dipole in a direction broadside to the dipole or, rarely, the dipole does better than the vertical in a direction in-line with the dipole wire. In an isolated case it appears the vertical angle of departure of a signal from the transmit antenna could very well influence the relative results of two different receiving antennas the same distance from the transmitter.

Indeed, all of the above has been known for a long time. However it is a propagation condition you should experience to fill in your practical knowledge of antenna results. If you are a one-band enthusiast you might consider two or three antennas that you can switch among as conditions change. Also you'll become more tolerant of the never-ending discussions of vertical versus horizontal, Yagi versus quad, etc. Most of all, enjoy yourself.

12-5. The Antenna Tuner

The antenna tuner adds versatility to your antenna installation(s) and does a big favor for your transmitter. When adjusted correctly it presents a proper load for your transmitter despite an antenna system mismatch. By so doing, good output power is sustained and the transmitter is not overheated or its components damaged. A tuner does not improve the performance of your antenna system; it does not make a bad antenna better. Up to a limit it does permit you to use a mediocre antenna to the extent of its capability without causing transmitter output troubles.

A tuner must be tuned. When you change bands it must be reset and tuned to the new band just like some transmitters. The same applies when you move your transmitter operation from one segment of an individual band to another. If the antenna system does match your transmitter there is no special benefit with the use of the tuner in terms of transmitter output adjustment.

How then does a tuner help? It is helpful when you wish to operate one antenna on several bands or when you like to move freely from one end of a band to another. This is especially true on the lower fre-

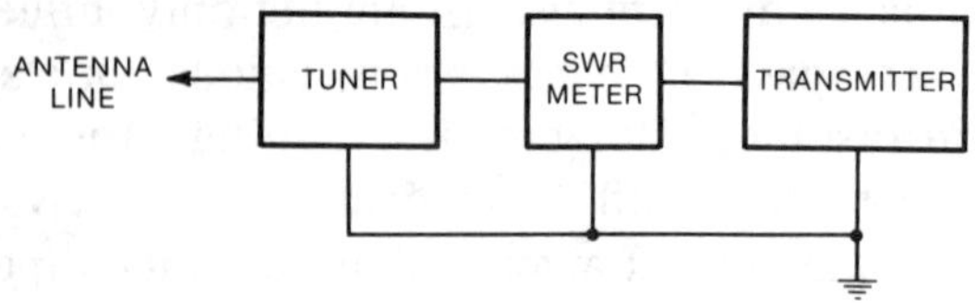

Fig. 12-6. Tuner location in ham setup.

quency bands and on the 10-meter band with its wide frequency span. These advantages will be discussed in connection with appropriate antennas that follow. In most situations a tuner is most beneficial when the degree of mismatch is troublesome but not excessive. This is usually the case for the previously mentioned modes of operation.

Although a tuner can often correct a serious mismatch it is sometimes of no great value because the antenna system itself is a very poor performer.

On transmit a good tuner can attenuate any harmonic output from the transmitter. On receive it helps to block off-frequency signals outside of its own bandwidth that might be strong enough to break through and overload the receiver input system.

A tuner can be a big help to the antenna experimenter or designer. You can check out new types as well as antenna ideas of your own without placing an unnecessary strain on your transmitter output circuitry. You can put your antenna ideas into action on the air despite a significant system mismatch. If performance appears to be good you can get down to the procedure of obtaining a good match. This happened to me in the design of the first low-frequency, single-loop triangle antenna back in the 1960s. At first it was expected that the antenna would have to be much longer in overall length and would need a 4-to-1 balun matcher. The antenna performance was good despite the mismatch. The tuner and SWR meter soon disclosed that no high-ratio balun was necessary and that resonance could be obtained with a shorter overall length as compared with the usual quad calculations. These changes resulted in a proper match to a 70-ohm line.

The tuner in most ham installations is located near the transmitter with the SWR meter positioned between tuner and transmitter (Fig. 12-6). Be certain to connect your station equipment ground to your tuner ground. This is especially important when using a single-wire, end-fed antenna. A convenient setup includes an SWR meter in the same unit such as the

Fig. 12-7. The MFJ-941D tuner and SWR meter combination. (Courtesy MFJ Enterprises, Inc.)

one shown in Fig. 12-7. Tuners usually have a number of other attractive features that provide more convenient operation and testing, such as coaxial switching between two coaxial inputs, capability of switching around the tuner connecting antenna directly to transmitter, high impedance single wire input, included 4-to-1 balun, power and SWR metering, dummy load capability and others.

12-6. Tuning a Tuner

Read the tuner instruction book!

A convenient approach to tuner adjustment is to first set up your transmitter for a 50-ohm resistive match using a dummy load. Prepare a table of settings for the key frequencies you are most likely to operate on if you do not wish to use a dummy load frequently. If you have no dummy load an alternative plan is to keep records of settings that you have checked out previously feeding into 50–70-ohm antenna systems.

Lacking the above information, set the transmitter tune and load adjustments according to dial-calibration information that is included in your transmitter instruction manual for 50–70-ohm matched settings together with your previous experience with the transmitter.

Usually you can make the initial tuner adjustments on receive. Listen to the background noise or a weak signal on or near your chosen tune-up frequency. Adjust the bandswitch and the tune controls of the tuner for maximum receive signal.

Put your transmitter on low tune-up power and adjust the tuner controls for minimum SWR. Now trim the transmitter tune and load controls for maximum power output. You may need to repeat the above two steps to obtain a minimum SWR reading. Now switch over to high transmitter output and retune again. It seems like a lot to do but you quickly adapt as you learn how your particular tuner and transmitter interact and how your tuner responds to various antenna types. Keep a record of proper settings for the various bands and antennas. It makes tune-up easy the next time.

Typically, when checking out a different antenna I use the antenna bridge to trim and bring the antenna close to desired resonance. Then I connect the transmission line and return to the radio room. The receiver is now used and the tuner controls are set according to previous experience with other antennas on the same frequency. Then the tuner is fine tuned for maximum signal or noise level. Next, I switch the transmitter to tune-up power and find out if the SWR can be brought down to a safe and minimum value for the transmitter by adjusting both the transmitter and the tuner. Finally, fine tuning is done with the transmitter on regular power. Usually only minor tuning changes need to be made.

If results are not acceptable on low-power operation it can be an indication of an antenna system problem with a mismatch so serious it cannot be accommodated. Remember also that despite its versatility and usefulness the tuner cannot improve upon a poor-performing antenna even though it sets up a perfect match to the transmitter output.

12-7. Tuner with 20-Meter Vertical Antenna

A valid application for a tuner can be demonstrated in connection with the 20-meter ground-plane vertical of Topics 12-3 and 12-4. The vertical is cut for resonance at 14,240 kHz. Insertion of a tuner ahead of the SWR meter made it possible to maintain a 1-to-1 SWR ratio at the transmitter from one end of the 20-meter band to the other. Only a slight resetting of the tuner controls is necessary when changing frequency. Tuners can be used with other basic antennas and provide valid wideband matching capability. Dependent upon the antenna type, another valid application permits a match on two or more bands with the same antenna. A valid tuner application refers to an assignment that adds versatility to a basically sound antenna. The above differ from a stop-gap use of a tuner for loading a less effective antenna which, at times, can also be advantageous. Such application does give you an operable antenna even though you should not expect too much from the system. A two-band valid application for a tuner is presented in Chapter 13. A variety of tuner use innovations are a part of Chapter 14.

Chapter 13

Antennas for the New Novice Operator

THE primary objective of the new novice operator is to prepare for the technician exam and then go on to the general class code and theory examinations. There are, however, important secondary objectives which are a definite aid in both preparing for the examination and gaining experience as a ham radio operator and technician. On-the-air operation aids in the preparation for the code test, provides technical experience, and gives direct experience that can apply to the written questions.

13-1. Introduction

The antenna is a part of both technical activity and operation. It is a much talked-about subject and some fundamental knowledge and experience makes conversation easier on the novice bands and eventually on the general band allocations.

Novice operation is a training experience. In addition, through it you often find your favorite niche in ham radio activities. So take advantage of this time. Try all bands allocated to the Novice Service.

This chapter concentrates on simple, low-cost and effective homebrew antennas. You can learn much and have fun. You can get a feel for new types of operation and propagation conditions. The antennas given in Fig. 13-1 will work well for operation on the novice bands. Please read Chapters 2 and 12 before you begin your antenna construction, tune-up, and on-the-air operation. Don't forget to heed the following warning you find on most bought antennas (Fig. 13-2). It is every bit as appropriate for homebrew antennas.

13-2. A Vertical Antenna for 15-Meter DXing

DXing on the 15-meter band is an exciting part of novice operation. The ground-plane quarter-wave vertical is a fine DX antenna. Also if you wish to have the benefit of a limited amount of broadside directivity the inverted-V performs very well. Dimensions for 15-meter operation are given in Fig. 13-1. Except for the lengths of the elements you can use the same construction details given in Chapter 12 for a 20-meter ground-plane vertical. Don't forget the additional construction guidance given in Chapter 2.

You may not wish to use an antenna bridge initially. If you construct the antenna carefully, your SWR meter readings are sufficient and will be satisfactory. If the minimum SWR at antenna resonance is higher than 1.5 to 1 you may wish to do a bit of antenna trimming. Details for doing so were given in Chapter 12. The performance of this simple low-cost antenna will surprise you. Use it as a start for your antenna experimentation.

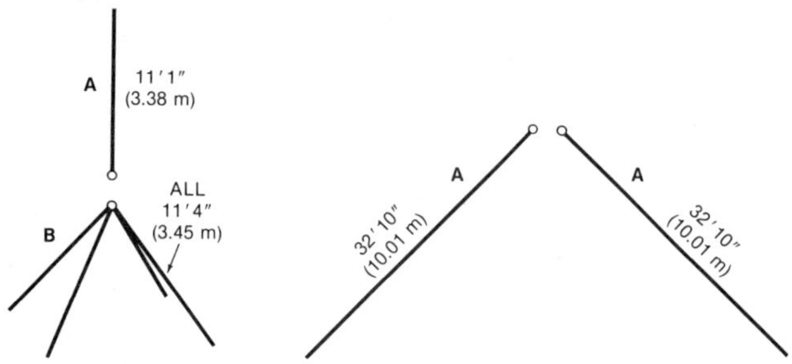

(A) 15-meter vertical. *(B) 40-meter inverted dipole.*

Meter Band	Freq. kHz	A λ/4	B Radial	C Director	D 0.12λ Spacing
80	3725	62′10″ (19.15 m)	64′5″ (19.63 m)	119′6″ (36.42 m)	36′1″ (11.0 m)
40	7125	32′10″ (10.01 m)	33′8″ (10.26 m)	62′6″ (19.05 m)	16′7″ (5.01 m)
15	21,150	11′1″ (3.38 m)	11′4″ (3.45 m)	21′1″ (6.43 m)	5′7″ (1.7 m)
10	28,150	8′4″ (2.54 m)	8′6″ (2.59 m)	15′10″ (4.83 m)	4′4″ (1.32 m)

(C) Dimensions.

Fig. 13-1. Basic novice antennas and dimensions.

Fig. 13-2. Warning label.

13-3. A 40-Meter Sloper Antenna

A sloper antenna is a slanted half-wave antenna with one end of the antenna attached to mast top and the other end near to ground level as shown in Fig. 13-3. It shows a modest directivity in the direction of its slope and considerably less signal pickup from its rear. Thus you can orient it in some favorite DX direction and squelch the QRM from the opposite direction a bit. This antenna has been a favorite arrangement of mine for many years and I have used it extensively for ham radio operation and SWB listening. I have used one on 160 meters to work Europe.

Construction and other details are to be found in Topic 2-7. For best directivity bring the lower end rather near to ground, approximately 5–8 ft (1.524–2.44 m). Inasmuch as the low end is near ground level, it is easy to swing the sloper around to other directions. Drive two or three additional stakes into the ground for easy directivity changeover. Nights you can cut back on European broadcast QRM by sloping the antenna in the opposite direction. Since the pattern of the sloper is rather broad, positioning is not overly important and it need not be oriented critically. The transmission line is attached using a plastic connector block with attached terminals or with a purchased dipole-to-coaxial connector as shown.

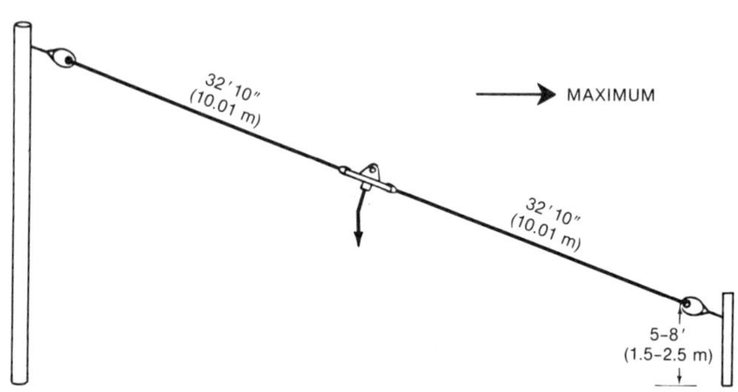

Fig. 13-3. A 40-meter sloper.

The sloper does give you some directivity experience and it is certainly much less expensive than a highly directional antenna. It provides a space-saving advantage as well.

13-4. Two-Element 15-Meter Inverted-V Beam

A two-element inverted-V beam has some gain, a reduced back pickup and lends itself to the construction of a low-cost beam. Details on this construction

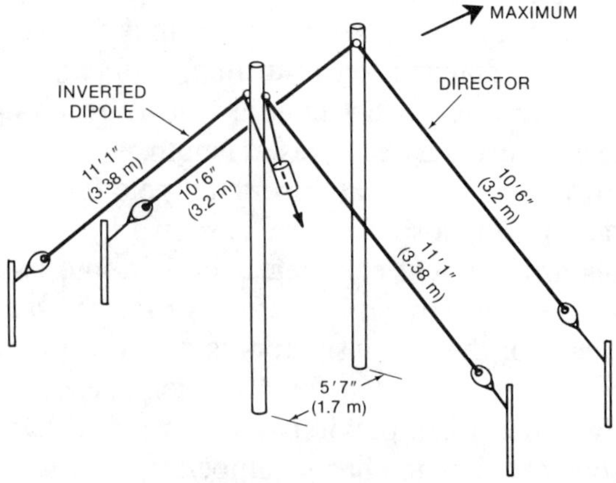

Fig. 13-4. A 15-meter two-element inverted-V beam.

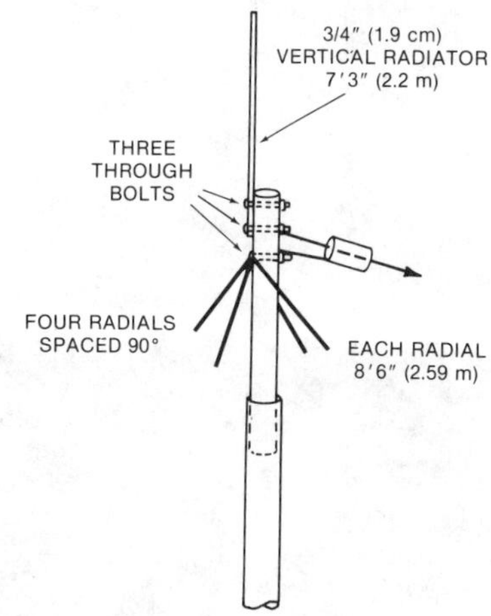

Fig. 13-5. A 10-meter elevated vertical.

were presented in Chapter 6. A 15-meter novice band version in Fig. 13-4 uses a parasitic element spaced about 0.12 λ in front of dipole. This spacing provides an adequate match to 50-ohm line. The spacing also provides a good reduction in back pickup which is often a help in reducing QRM. Remember, the beam directivity and gain can be improved somewhat by making the elements more horizontal by mounting the end stakes farther from the mast and increasing the length of the support ropes between the end insulators and the masts.

In my location the front-to-back ratio is often more useful than forward gain on 15 meters. When the path is open to Europe, the problem is often not the weakness of European signals but the severe QRM from stateside signals. Remember this some day when you become an active sideband operator. A good pair of antennas for operation is this fixed beam and the 15-meter vertical with separate lines and a two-position coaxial switch.

13-5. A 10-Meter Elevated Ground Plane Vertical Antenna

The ground-plane quarter-wave vertical antenna is especially effective on 10 meters. It is an omnidirectional antenna and capable of handling signals regardless of the direction of the DX opening. The radiator itself is short in length and lends itself to high mounting. The ground plane vertical responds well to the low angle DX signal arrival and also performs well for local mobile reception. You might think about organizing a local novice 10-meter network so you become familiar with 10-meter operation and become ready for the 10 meters that soon will become an amazing DX-active band as the sunspot cycle rises.

The vertical of Fig. 13-5 consists of a length of 3/4-in aluminum tubing held to the top of a PVC mast with two through bolts. One bolt is mounted at the very bottom of the aluminum tubing that forms the vertical radiator and also acts as the terminal to which the inner conductor of the coaxial line is attached as shown in Fig. 13-6. The bolt/nut terminal immediately below the end of the tubing serves as a terminal for the four wire radials. The braid of the coaxial line connects to this terminal. It is a very simple construction. The vertical radiator is 7 ft 3 in (2.2 m) in length. This is considerably shorter than the table value. However, the radiator is 3/4-in (1.9 cm) in diameter and shows more end effect than a small diameter piece of wire. Also the two through bolts have an influence on resonant length. Vertical antennas will do well on the 28.300 to 28.500 MHz novice sideband spectrum.

13-6. Two-Band Inverted Dipole Antennas

Inverted dipoles do well on both the 40- and 80-meter novice bands. The inverted dipole is more omnidirectional than a straight dipole. Use the same construction detail offered in Chapter 12. Also refer to Chapter 2. Some minor end trimming may be necessary if you

Fig. 13-6. Vertical tube, terminals, and radials.

wish to place the very lowest SWR point in the novice band. The inverted dipole style makes it easy to release one antenna leg at a time, dropping the end insulator to a level that permits you to reach up and make any necessary changes in length.

It is easy to plan an inverted dipole for two-band operation using a single mast as shown in Fig. 13-7A. In this arrangement each leg is broken into two segments separated by an insulator. A jumper associated with each of the two insulators is closed for 80-meter operation and left open for 40-meter operation. This idea was described previously in Chapter 3. Drop the legs one at a time to change jumpers.

I went through the tune-up procedure very carefully to derive the dimensions shown in Fig. 13-7A. If your construction is close to these values you will have no trouble. An SWR meter can check out the results although it is not likely that there will be any significant trimming required.

An attractive arrangement that does not require any band change jumpers involves the cross-dipole plan of Fig. 13-7B. The two dipoles are connected to the same transmission line. Additional construction details were given previously in Chapter 5.

There is very little interaction between dipoles when they are erected at right angles to each other.

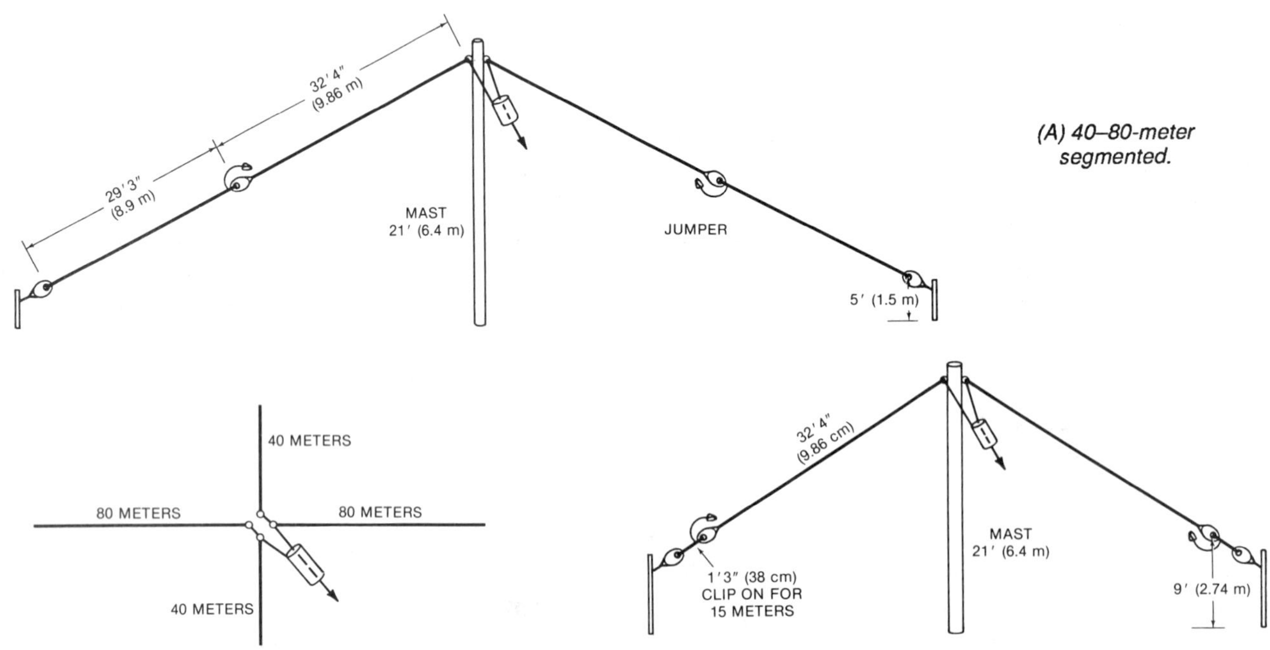

(B) 40–80-meter crossed. *(C) 15–40-meter clip-on.*

Fig. 13-7. Two-band inverted dipoles.

However if space is not available you can bring them considerably nearer to each other without any significant change in performance. However you may have to do a bit of trimming to keep minimum SWR in the novice bands. This is usually quite unimportant in terms of performance if the SWR does not exceed 2 to 1. It is a good idea however to try to bring it down to no more than about 1.6 to 1.

You can add 15-meter capability to a 40-meter dipole by adding a short end segment to each leg that can be clipped on or off depending upon the band of operation desired. The arrangement of Fig. 13-7C works out well for the 40-meter inverted dipole. Such an arrangement operates as a fundamental inverted dipole on 40 meters and a 3/2-wavelength dipole on 15 meters. For 15-meter operation each leg of the dipole is 3/4 wavelength long (a bit longer than a 40-meter quarter wave). Consequently, there is a low impedance at the point where the transmission line is connected. The basic concept of operating antennas on odd-harmonics of the fundamental cut of the dipole were presented in Chapters 5 and 6.

I worked out the dimensions carefully for this mode of operation. If you follow the dimensions given in Fig. 13-7C with reasonable care you will have a convenient and effective two-band antenna. On 15 meters this antenna has some good low-angle directivity off the wire ends. Consequently, you may wish to line up the line of the antenna wire in some favorite direction. However the 3/2 wavelength antenna if mounted in the inverted style is also reasonably omnidirectional and gives you two-band capability with ease. When 15 meters is active it is a simple matter to make the two jumper changes to enjoy a bit of DXing.

13-7. Tuner and 15/40-Meter Dipole Antenna

In Topic 13-6 the two-band inverted dipole used jumpers to change over between half-wave 40-meter dipole and 3/2 wavelength 15-meter operation. The use of the tuner eliminates the need for the jumpers. Instead, the changeover is made at the operating position by adjusting the tuner.

The leg length of the dipole can be either the 15-meter or the 40-meter value, or somewhere in between. Most often with this and similar antennas it seems to work best by cutting the dipole to the higher frequency. Here, it is 15 meters. The tuner easily loads up the antenna on 40 meters because the added physical length required represents only a short electrical length on 40 meters. There is little or no difference in performance with the added length as compared with an exact 40-meter cut. It is indeed a truly valid application for a tuner. The next topic describes how a random length of wire and a tuner can be made to perform well when the length of the wire is chosen wisely.

13-8. Random Wire Antenna with Tuner for the Novice Bands

A length of antenna wire attached to the single-wire input of a tuner can function as an effective antenna when it is of *proper length.* The preferred method of operation occurs when the wire is one quarter wavelength long or some odd multiple of a quarter wavelength. There are two advantages of this mode of operation. First, the quarter wavelength of wire is an efficient radiator of radio-frequency energy. Second, the low impedance presented at the tuner prevents stray high-level radio-frequency components in the radio room.

An effective random wire arrangement for 80 meters is shown in Fig. 13-8A. The overall length of the wire is 62 ft 10 in (19.15 cm), only one half the length of a full 80-meter dipole. Thus the random wire with tuner is attractive for use when you don't have the space to stretch out a longer antenna. It is not

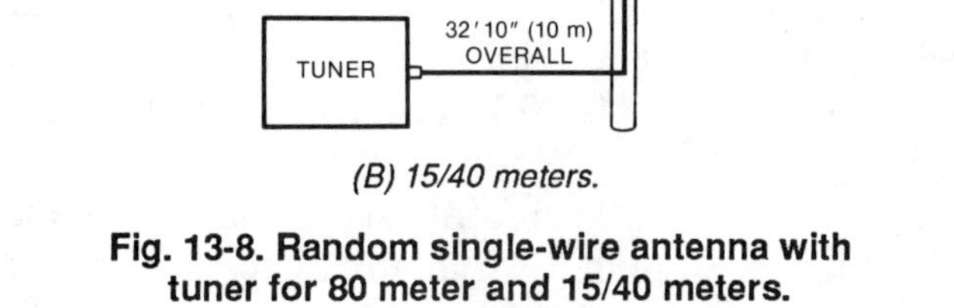

Fig. 13-8. Random single-wire antenna with tuner for 80 meter and 15/40 meters.

Fig. 13-9. Austin triband vertical antenna for 2 meters, 1¼ meters, and 70 centimeters.

quite as good as a full length dipole but you can get out quite well! The wire need not be in a straight line. Stretch it out as much as you can. However, it is better not to fold it back upon itself. If space is so limited that you cannot erect a 40-meter dipole, try a 32 ft 10 in (10 m) random wire with tuner.

In planning any random-wire antenna the length of the wire is measured from the far end of the wire to the single-wire input point of the tuner. Look at the simple 15/40 installation positioned close to a window of the radio room in Fig. 13-8B. The single antenna wire begins at the top of an 18 ft 6 in (5.64 m) PVC mast and ends 32 ft 10 in (10 m) later at the tuner input.

An attractive feature of this installation is that the 40-meter length also approximates a ¾-wavelength span on the 15-meter band. As a result you also

Fig. 13-10. The Cushcraft 1¼-meter Yagi.

have a simple arrangement that works on both bands. More data about odd-multiple longwires and how to choose the most favorable length for a single-wire antenna is presented in Chapter 14. There are a number of antenna types in Chapter 14 you may wish to try. Adapt them to 10-, 15-, 40-, and 80-meter dimensions appropriate for the novice frequencies.

13-9. 1¼-Meter VHF Antennas

The 1¼-meter (220 MHz) band is a popular one with novices because phone operation is permitted. Antennas are small and basic, and it is usually more economical to buy than build. A list of antenna sources and manufacturers is given in the Appendix.

If you plan to install a mobile or fixed antenna for this band it would be advisable to first go over Chapter 15. Also there are dimensions given for this band in Chapter 16. A review of Chapter 10 may also be helpful.

Read the descriptions associated with Figs. 15-1 and 15-2. Shorter length versions of these two 2-meter antennas are available for 1¼-meter operations. Also the antenna of Fig. 15-4 can be cut for 1¼ meters. Read Topics 15-1 through 15-3.

If you are a novice you are no doubt working toward a technician grade. The antenna of Fig. 13-9 is worth a thought because it is a triband model by Austin. It provides a low SWR on 2 and 1¼ meters and 70 centimeters. Hence, with one installation you are ready for operation on two adjacent bands when your technician license comes through. The 2-meter band is an especially active one. This antenna is available for mobile or fixed mounting. Refer to Topic 15-8 which discusses a duoband version and its mobile/fixed capability.

You may wish to try a beam antenna such as the Cushcraft seven-element Yagi of Fig. 13-10. This antenna has seven elements and a beam length of 5 ft 9 in. The gain is 10.5 dB with a good directional pattern. Often you can aim this antenna on some favorite repeater which is some distance away and continue to do well on your close-by repeaters. A more suitable arrangement is to use both an omnidirectional and a beam antenna. Run separate lines to a coaxial switch on the radio room.

You may wish to mount your beam on a rotator or use the simple arm-power arrangement of Topic 15-6. A beam is useful for contesting on simplex frequencies. However you can do contesting without a rotator and, for routine repeater operation, a rotator is really not required. Later, if you get around to avid DXing on sideband, a beam and rotator can be of help.

Presently two meters is the most active VHF/UHF band. Most technician licensees begin their operations on this band. Many current and newly licensed novices will begin their ham operations on 1¼ meters. When they obtain their technician or general license they can begin their operations immediately on 2 meters if they purchase a duobander antenna initially. An initial tribander would give them 2 and 1¼ meters, and 70 centimeters capability. Actually the opening of 1¼ meters is likely to make the 2 meters and 1¼ meters the two most active bands followed closely by 70 centimeters.

Austin has a variety of combinations with appearances similar to those shown in Figs. 13-9, 15-22, and 15-25. More are likely to appear.

Chapter 14

Antennas for General- and Advanced-Class Licensees

REGARDLESS of your class of license—general or advanced—you can have a whole backyard of enjoyment homebrewing cheap and effective wire antennas. The new or more experienced general or the advanced licensee can use any of the antennas described in Chapters 12 and 13 but they must be cut a bit more carefully for the band segment you desire. Use the dimension tables of Chapter 16 to find the wire lengths required for the band(s) or segment of a band that appeals to you. This is the fun way to go and it will be a profound experience and a special satisfaction to learn how much you can do with so little in the way of antenna materials and, at the same time, erect an efficient and effective antenna.

14-1. Introduction

You can work the world with an inverted dipole or one of the other basic antennas or variations thereof. It is an achievement you should not miss and it will make you a wiser and more understanding, considerate, and better operator. This metamorphosis cannot be expressed in words; you must experience it yourself. Actually it may make you a lifetime ham operator rather than a bored drop out.

You will find the descriptions that follow become more meaningful if you first go over some of the earlier chapters, especially 1, 2, 3, 12, and 13. Other chapters also contain background with a direct bearing on the antennas that are covered for ham operations.

A variety of unique arrangements are described with special emphasis given to antennas that operate on more than one band, can be erected in a small area, have some gain or a special pattern, and need not be sky scrapers.

14-2. Segmented Dipole Antennas

The segmented dipole is practically a no cost method of obtaining two or three band operation from the same antenna on the low-frequency HF bands. Of course you may need to take a short walk to the antenna to make the necessary jumper or clip-on changes when shifting bands or between sections of one band. The inverted-V dipole is the preferred arrangement because its wire ends are supported at near ground level. However, a suitable lanyard arrangement works out well with a regular horizontal dipole. New general class operators can use this plan to pick up some operating time on a number of bands as they leave novice operations and enter this new vista of operating frequencies.

The new general operator can also pick up some good sideband experience on 40 meters. It is active all day and into the early evening before the foreign

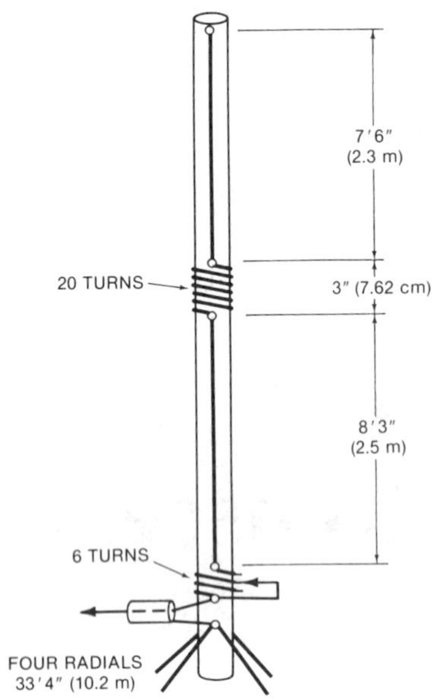

Fig. 14-1. Center-loaded 40-meter vertical antenna.

broadcast stations make communication difficult, especially in the eastern part of the USA. A short length of line that you can clip on the end of each leg of the 40-meter dipole will give you 15-meter capability. Refer to Fig. 13-7C. The general segment of the 40-meter band suggests dipole legs of 32 ft (9.75 m) and a clip-on segment for 15-meter operation of 1 ft. In our sideband version the 40-meter dipole length is 32 ft 1 in (9.78 m) with a 1 ft 2 in (35.6 cm) clip-on segment for 15-meter operation. In some installations one need not be overly critical about dimensioning. The figures given established a very minimal SWR at the specified frequencies. Keep in mind however that with low height and very long antennas there is often a considerable amount of trimming required. By now you are probably well adept at making the necessary trimming to obtain good SWR operating conditions.

The same clip-on method can also be used to obtain a more favorable SWR on either end of the phone or CW segment of the 40-meter band, and it is especially useful on the 80- and 160-meter bands.

Typically an inverted dipole is cut for the phone segment of a band. This portion is most often at the high frequency end of the band assignment. When you wish to operate CW, attach a small clip-on to each leg and you can obtain an optimum low SWR reading at the CW end of the particular band. Typical clip-on values for 40- and 80-meters are 8 in (20.3 cm) and 3 ft 5 in (1.04 m), respectively.

A 40–80 meter jumper plan for changing bands was first presented in Fig. 13-7A. In this arrangement you free one leg of the dipole at a time to make the band change. Typical values for the antenna plan of Fig. 13-7A for setting the minimum SWR near 7200 on 40 meters and 3825 on 75 meters are:

7200–32 ft 1 in (9.78 m)
3825–32 ft 1 in plus 28 ft 3 in clip-on = 60 ft 4 in (18.39 m)

Considerable trimming was required because the completed antenna was only 21 ft (6.4 m) at its apex and it extended down to a wire end height of only about 6 ft (1.83 m). Of course all these antennas can be mounted considerably higher and must be trimmed carefully for best operation at whatever height is used. However, before going to all this trouble, gain some experience with antennas in a much easier way and also be amazed by how well a low antenna performs.

14-3. Center-Loaded 20-Ft 40-Meter Vertical Antenna

The center-loaded vertical antenna has been a popular installation for a long time for VHF/UHF operation. The very same technique does well on the HF bands too. An example is the 40-meter version of Fig. 14-1. In this plan a loading coil is positioned just above the center of the PVC mast (Fig. 14-2). The center loading coil wire is made of the same wire as the antenna and consists of 20 turns of plastic-covered No. 16 wire wound around the 1½-in (3.81 cm) ID mast itself. The lengths are given in Fig. 14-1. The vertical length and turns of the center loading coil are selected to produce resonance at the high-frequency end of the band.

At the base of the antenna a six-turn loading coil (with taps) is included. You can use these turns to resonate the vertical for very minimum SWR at a preferred spot in the 40-meter band. Four 33 ft 4 in (10.2 m) radials complete the vertical. They are run down the mast a short distance and fan out at 90° just beneath the ground surface. Use the same *insulated* wire you did for the vertical. The wires act as ground plane resonant radials and can be placed on the surface or just beneath it.

The mast is 18½-ft (5.64 m) overall but is elevated about 2½ ft (.762 m) above ground, putting the top of the vertical at 21 ft (6.4 m) above ground.

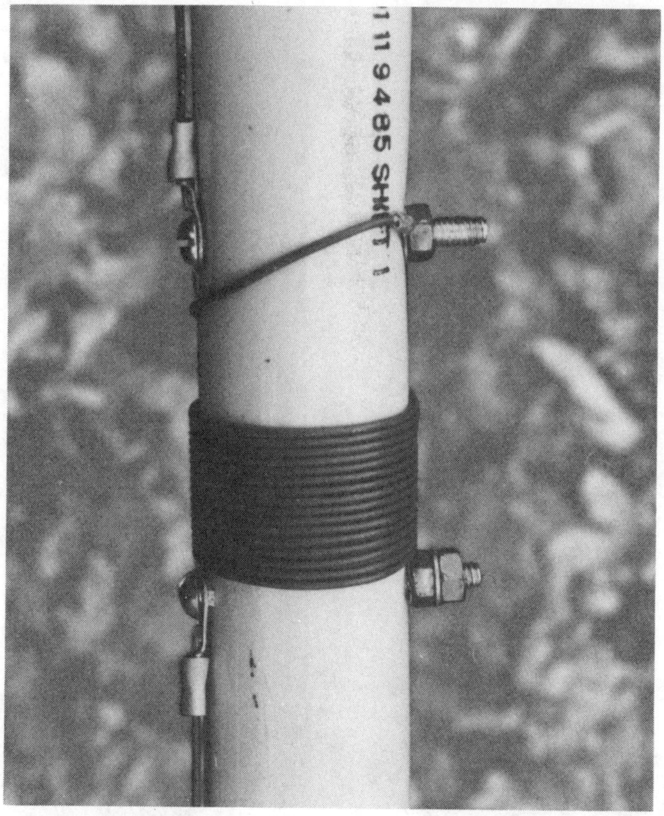

Fig. 14-2. A 40-meter center-loading coil.

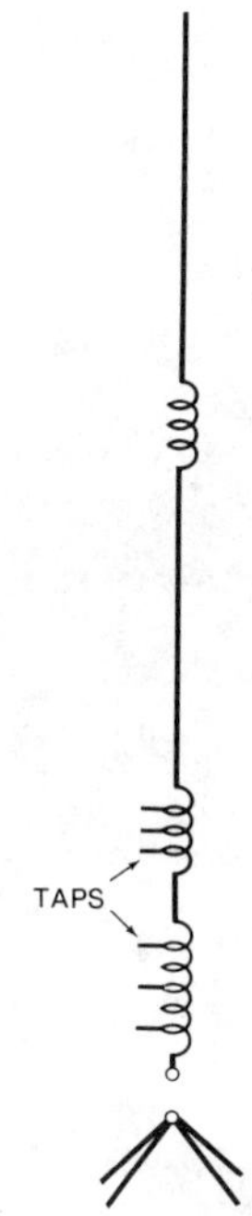

Fig. 14-3. A 40- and 80-meter loaded vertical.

14-4. The 40- and 80-Meter Loaded-Vertical Antenna

The center-loaded 40-meter vertical antenna of Topic 14-3 can be expanded to include 80-meter operation with the addition of a multiturn base coil as shown in Figs. 14-3 and 14-4. For fine tuning, the top base coil consists of 29 turns close wound and tapped each 16 in (40.6 cm) along the length of the wire. The lower base coil comprises 23 turns with taps each 5 ft along the wire. The actual tap selection is made with a length of line and its end connector clip that extends from the top coaxial terminal to the top of the upper base loading coil. A weatherproof connector tip can be seen in Fig. 14-4. In our installation resonance on 7200 kHz occurred with the connector clip attached to the third tap down from the top of the top base coil. For 75-meter resonance on 3800 kHz the connector clip was attached to the first tap from the top of the lower base coil. The various taps permit you to choose a preferred minimum SWR frequency.

The vertical antenna performs very well on 80 meters using only the four 40-meter radials. If space is available, you may wish to try some additional 75-meter radials. I worked into Poland with three 75-meter radials attached.

14-5. A 160-Meter Loading Coil for 40- and 80-Meter Vertical Antenna

The vertical antenna described in Topic 14-4 and its coils perform well on 40 and 80 meters using the four 40-meter insulated radials pushed just under the ground. An additional outrigger coil (Fig. 14-5) can be used to put you on 160 meters. The coil has 82 turns close wound on an 8½-in (2.59-m) length of 2 in ID pipe. Wire it in series between the bottom of 80-meter base-loading coil and the top transmission line terminal as shown in Fig. 14-6. Connect a jumper across the coil for 40- and 80-meter operation.The antenna will load using the 40-meter radial system and the performance will surprise you. Not much space is required considering its capabilities on the three lowest frequency ham bands. You will have a fine 40–80–160 reference antenna.

If space is available you can go for a trio of 80-meter and 160-meter radials. Again use resonant radials (insulated wire) stretched out on the ground surface or just beneath and you will avoid some of the variables of an earth ground.

(A) 40 meter.

(B) 80 meter.

Fig. 14-4. Base coil taps for operation on 40- and 80-meter band segments.

14-6. A 40-, 80-, and 160-Meter Transmit Capability with All-Band Reference Antenna

If you constructed the all-band reference receiving antenna detailed in Topics 9-2, 9-3, and 4-6 you already have capability for 40-, 80-, and 160-meter ham-transmit mode. No change is needed to operate on 40 meters. Choose the tap point carefully to load the antenna in the desired section of band. For 80- and 160-meter loading you must add one or two 80-meter radials. These radials allow proper loading on all three bands. As mentioned previously, this vertical has been erected just outside the radio room window. Only 20-ft (6.1-m) of coaxial cable provides a link to any receiver or rig. It is hard to beat as a reference installation on SWB, ham, and other HF bands, plus it even works the broadcast band and LW frequencies.

14-7. Close-Spaced Inverted Director for 40 Meters

An inverted dipole with a close-spaced inverted director does a fine job on 40 meters and other lower frequency bands if cut to appropriate length. The spacing requirement is attractive because it is only 0.1 wavelength and will not require too much space at the

Fig. 14-5. Outrigger attachment of 160-meter loading coil.

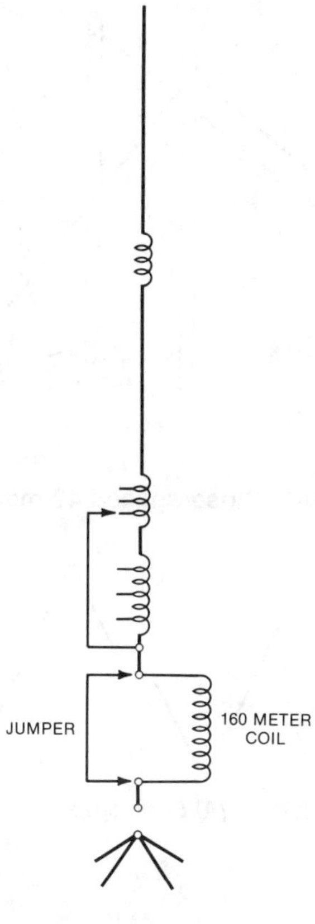

Fig. 14-6. Outrigger loading coil connections.

mounting site. On 40 meters 0.1 wavelength is approximately 13–14 ft as shown in Fig. 14-7. In the practical tests made here the beam showed a wide angle of radiation in the forward direction. There was a modest gain as is to be expected with only one additional parasitic element. The gain and signal pickup was reduced to the rear and from received reports appeared to be most pronounced in a narrow angle directly to the rear. If you would like a small amount of forward gain and some back rejection give this antenna a try. The overall benefit can be significant because a limited gain and limited rejection can result in a significant improvement when you are troubled with QRM from the rear.

14-8. The Basic Horizontal Zigzag Dipole

The horizontal zigzag construction permits you to erect an antenna on a small lot despite the long lengths of wire required by a low-frequency dipole. There are various configurations that produce good results from the basic zigzag of Fig. 14-8A to the end-fold arrangement of Fig. 14-8C. Much depends upon your own antenna site. Stretch out the antenna arrangement as much as you can and minimize the amount of fold back, although a complete fold such as Fig. 14-8D does produce acceptable results when proper matching is established. The best plan is to extend your antenna as much as your site permits.

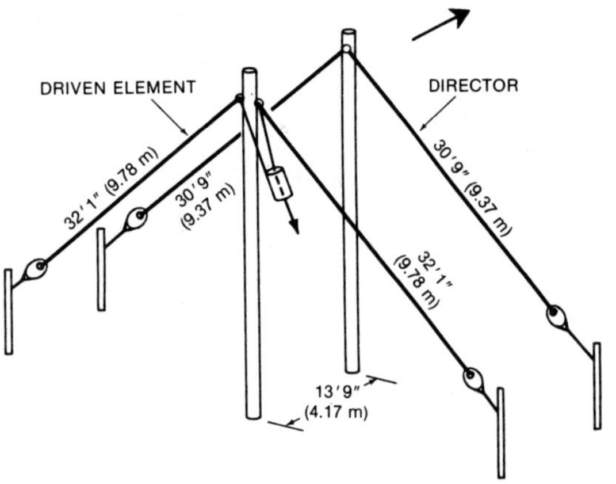

Fig. 14-7. Close-spaced 40-meter beam.

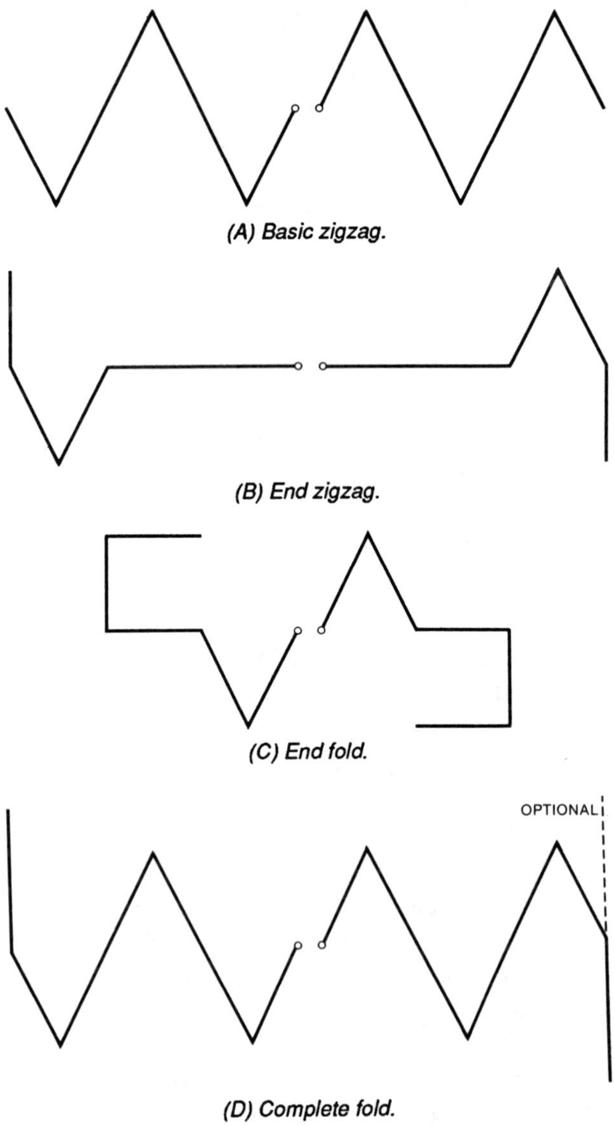

(A) Basic zigzag.

(B) End zigzag.

(C) End fold.

(D) Complete fold.

Fig. 14-8. Zigzag dipole plans.

Keep the ends separated as much as you can. Better all-around performance on 160 meters with Fig. 14-8D seems to favor the ends angled in opposite directions rather than in the same direction as represented in the optional arrangement.

A disadvantage of the zigzag plan is that additional masts are required for support. A savings in support structure is possible if you employ an inverted-V construction tapering down to lower and lower masts and finally terminating the two ends at fence-post level.

The plan of Fig. 14-9 works out very well for a two-band arrangement. The 40-meter antenna is an inverted dipole with its ends at approximately the 10-ft (3.05 m) level as shown in Fig. 14-9A. The mast height was approximately 22 ft (6.71 m). The zigzag added sections that provide 75-meter operation are at 10-ft level. Two jumpers are included at the 10-ft level and permit easy band changeover using a 5-ft stepladder.

The ends zigzag in opposite directions. Stretch out the triangular ends as much as you can for best results. However the antenna still performs well if there is a complete 90° L-shape at each end. This configuration would be two end legs of 29 ft 9 in (14 ft 11 in plus 14 ft 10 in) but angled 180° to each other as shown in the arrangement of Fig. 14-9C. Use this arrangement only when you're tight for space and wish to load up with ease on 75 meters.

14-9. A 40-, 75-, and 160-Meter Zigzag Dipole

The arrangement of Fig. 14-10 works well and permits good 160-meter performance in approximately a 100-ft (30.5-m) square area. This antenna is an extension of the basic arrangement of Fig. 14-9. The 40-meter segment is again an inverted dipole. The ends of the inverted dipole drop down to the 10-ft (3.05-m) level, and the remainder of the antenna remains at the same level with appropriate jumpers that can be reached for frequency changing among the three bands.

In my own installation the inverted dipole showed minimum SWR at 7200 kHz. When an additional 29 ft 9 in (9.07 m) segment is jumped into the circuit we are able to locate our minimum SWR at approximately 3820 kHz on the 75-meter band. When

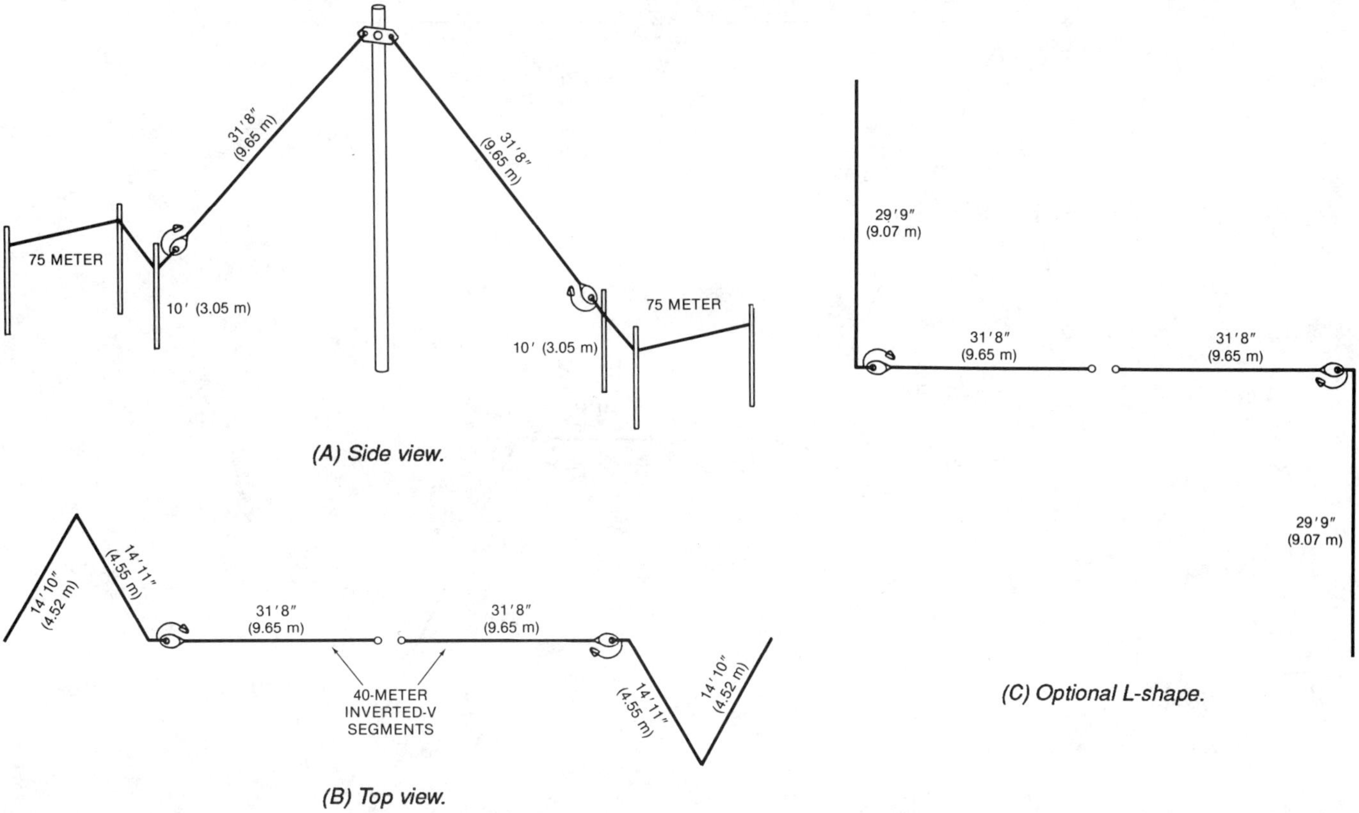

Fig. 14-9. A 40-meter inverted-V with 75-meter zigzag dipole.

the extra 63 ft 6 in (19.35 m) segments were jumped we were able to position the resonant frequency of the antenna on 1950 kHz. Finally adding 6 ft 3 in (1.9 m) to each end reduced the 160-meter band frequency to approximately 1855 kHz. Thus, operation on either sideband segment of the 160-meter band is possible. In fact the ends of the antenna taper down to the 5 ft (1.5 m) level and the 160-meter band section changeover can be made at ground level. Only four additional 10-ft (3.05-m) lengths of PVC piping permitted this two band addition to the 40-meter inverted dipole. Two metal fence stakes and short lengths of rope supported the ends of the antenna.

14-10. Tuner for Vertical and Dipole Antenna Variations

Tuners perform well with many of the antennas described previously in this chapter. For dipoles and ground-plane verticals they permit you to trim and set your antenna for some favorite spot in a given band and maintain a 1-to-1 ratio when you decide to move your operating frequency to some other part of the band. The MFJ Enterprises 941D tuner allows freedom of movement in the band, for even the zigzag configurations.

All the loaded verticals work well with the tuner and make the selection of an appropriate tap much less critical. On 40 meters there is no problem in maintaining a 1-to-1 SWR ratio over the entire band. Often a single tap maintained a reasonable match over much of the 80- and 160-meter bands. You can do a bit better with the wise use of two or three taps of adequate frequency separation on each band.

If you can't get the SWR below 1.5 to 1 it is usually an indication that your antenna matching system is not working effectively. There is probably such a high SWR on the line that it cannot be resolved by the tuner. Also there can be a lot of unwanted radiation around the radio room that it subtracts from the power that should be radiated by the antenna. Check your tap positions or jumper connections again. Then reset the tuner and keep records of the proper settings.

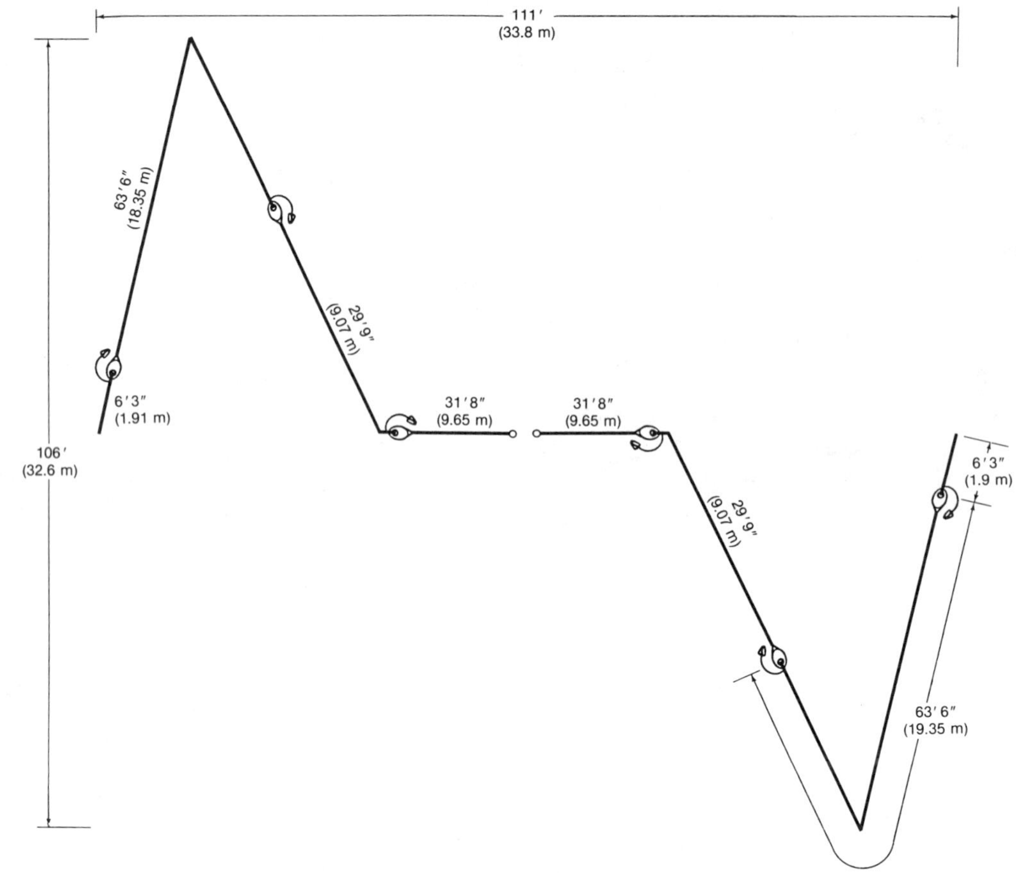

Fig. 14-10. A 40-75-160 zigzag.

14-11. Single Longwire Antenna with Tuner

A single longwire antenna is an inexpensive way of obtaining gain and directivity and, with a tuner, multiband operation. Good results are obtained when the longwire is made an odd multiple of a quarter wavelength. By so doing a low resistive impedance is reflected right to the input of the tuner. Odd quarter wavelengths for various frequencies related to single sideband portions of the bands (except for 30 meters) are given in Table 14-1. As with other tables, the top figure gives the dimensions in feet (and inches) and the bottom figure is the dimension in meters. Should you want to calculate for other frequencies, the constants are given at the bottom of the table. The longer (in wavelengths) the single wire antenna, the higher is its gain. Directivity of the longwire antenna is off the end of the wire.

When erecting a single wire, cut the length to the value given in the table. Usually you will have to trim the wire to obtain a minimum SWR on your specific operating frequency. In most cases you can trim back and obtain a good impedance match at the transmitter. Use the antenna bridge at the transmitter, measuring the single wire against station ground. Unless the antenna is quite high, the resonant length is usually 2 to 4% less than the table values. Often you can cut a 3/4 λ longwire antenna for direct match to the transmitter without the use of a tuner.

When you use a tuner, however, you can tune up to other bands especially those bands that also exhibit a multiple of an odd quarter wavelength. Some typical combinations are given in Table 14-2.

I tried out the first, third, and fourth combinations in Table 14-2 with great success. No stray RF or transmitter instability was present. Also I could tune up to some of the other bands with good results using these same combinations.

A tuner will indeed load the transmitter on some of the other bands using an antenna that is not odd harmonically related. That's fine. However, on some bands, usually on the higher frequency bands or on bands for which the longwire is near an even-harmonic length, there can be stray RF in the radio room which can be felt when you touch the metal of the microphone. Cross these bands off the list for this antenna. One thing you should not do with a tuner is tune up on an antenna length that is shorter than 1/4 wavelength on the operating band. The results are often erratic and/or poor.

I had fine results using the antenna in Fig. 14-11 that is 5/4 wavelength long on 40 meters. The dimension according to Table 14-1 is 169 ft. I dropped off about 6 ft to put the 5/4 wavelength resonant point in the 40-meter band, aided by the antenna bridge. A final dimension of 163 ft did well on 20, 12, and 10 meters; each band has a reasonably odd harmonic relation to this length. The tuner also loaded well on 75 meters.

One of the tuners was troublesome on 160 meters and could not resolve a match better than a 1.6-to-1 SWR. Actually the 163-ft length reflected a considerable reactance to the tuner that could not be resolved by the component values of the particular tuner. Both 10 and 160 meters are often a problem with a tuner when you try to match an end-fed wire that is not an odd multiple because these two bands represent the band limits of tuner design.

Table 14-2. Typical Antenna Combinations

Meter Bands	Multiple	Typical Length
40, 15	1/4, 3/4	34'
40, 15	3/4, 9/4	103'5"
75, 10	1/4, 7/4	60'
20, 15, 10	5/4, 7/4, 9/4	77'
30, 12	3/4, 7/4	69'
20, 12	3/4, 5/4	49'
40, 12	1/4, 3/4	30'

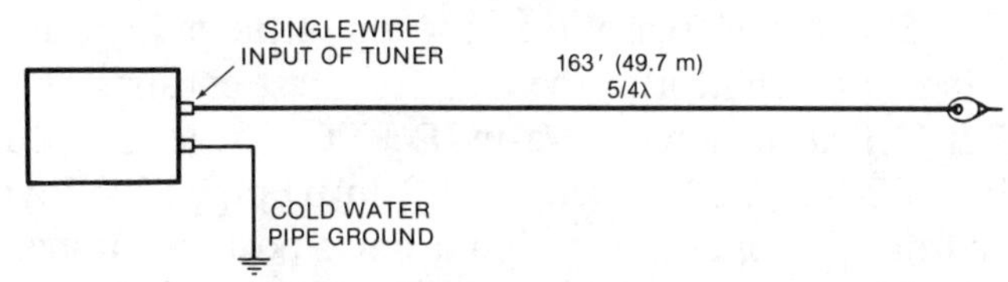

Fig. 14-11. Single-wire longwire with tuner.

The results on 15 meters were dismal with RF all about. I could touch the microphone case and change transmitter plate current and cause the SWR to rise over 3 to 1. This should be expected because the length of the longwire is almost exactly 16/4 wavelength at 21.3 MHz. An even multiple reflects a high impedance to the tuner and it became hot with RF voltage (physically hot, too). The answer was to cross 15 meters off the list for the 163-ft antenna.

Table 14-1. Odd-Multiple Dimensions for the HF Amateur Bands

Freq. (MHz)	1/4 λ	3/4 λ	5/4 λ	7/4 λ	9/4 λ	11/4 λ	13/4 λ	15/4 λ	17/4 λ	19/4 λ
1.85	126'	392'								
	38.5	120								
3.8	61'7"	191'	320'6"							
	18.8	58.2	97.7							
7.2	32'6"	101'	169'	237'6"	306	374				
	9.9	30.7	51.6	72.4	93.3	114				
10.12	23'1"	71'9"	120'4"	169'	217'7"	266'2"	314'10"	363'5"		
	7.05	21.9	36.7	51.5	66.4	81.1	96.0	110.8		
14.20	16'6"	51'2"	85'9"	120'5"	155'1"	189'9"	224'4"	259'	293'8"	328'4"
	5.02	15.6	26.1	36.7	47.3	57.8	68.4	79.0	89.5	100.1
21.30	11'	34'1"	57'2"	80'3"	103'5"	126'6"	149'7"	172'8"	195'9"	218'10"
	3.34	10.4	17.4	24.5	31.5	38.5	45.6	52.6	59.7	66.7
24.93	9'5"	29'1"	48'10"	68'7"	88'4"	108'1"	127'10"	147'6"	167'3"	187'
	2.86	8.9	14.9	20.9	26.9	32.9	39.0	44.0	50.0	57.0
28.5	8'2"	25'6"	42'9"	60'	77'3 "	94'6"	111'10"	129'1"	146'4"	163'7"
	2.5	7.77	13.03	18.3	23.6	28.8	34.1	39.3	44.6	49.9
Constants										
Feet	234	726	1218	1710	2202	2694	3186	3678	4170	4462
Meters	71.3	221.3	371.3	521.2	671.5	821.1	971.1	1121.1	1271	1421

14-12. 3/2 λ 20-Meter and 1/2 λ 75-Meter Sloper Antenna

The sloper antenna has gain in the direction of its slope. It isn't much but it is at a low wave angle—a big help for DXing. In addition there is reduced back response. On 20 meters there is some added gain because of its length. Each leg is 3/4 λ long and there is a low impedance at the feed point that permits a good match to 50–70 ohm line.

Start the sloper at the very top of the mast. I use a lanyard to pull it to the top. The mast consists of a 7-ft (2.1 m) length of 1 1/2-in ID PVC pipe telescoped into a 20-ft (6.09 m) length of 2-in ID pipe. The low end of the sloper is roped to a fence post, positioned to hold the antenna end 5 ft (1.5 m) above ground as shown in Fig. 14-12A. As shown, each leg is 51 ft 2 in (15.6 m) long to attain resonance at 7.2 MHz. Reasonable 75-meter results can be obtained by using a tuner to establish resonance on the 75-meter band.

By adding 11 ft 5 in (3.5 m) to each leg and moving the end fence post farther from mast (Fig. 14-12B) you can set up a 75-meter sloper. This can be done easily by dropping the sloper high end down with the lanyard. Do the necessary trimming to set resonance on 3.8 MHz or some other 75-meter frequency. A tuner will load this sloper very well on 20 meters.

If more omnidirectional results are desired the combination can be erected in inverted-V fashion as shown in Fig. 14-12C. In this configuration both ends will be near ground level and jumpers can be used to change over between bands.

A bit of trimming may be needed to hit an exact minimum SWR frequency for each of the three antennas. There is some difference in performance between the two slopers. The 20-meter sloper when tuned up to 75 meters caused about one S-unit drop as compared to a full half wavelength antenna on 75 meters. The situation is about the same for 20 meters when a full half wavelength sloper cut to 75 meters is tuned up on 20 meters. The difference shows up in the direction of the slope. Differences in other directions are more erratic and difficult to determine.

A tuner can also be used to tune up the antennas on other frequencies. Forget any band that causes RF to invade the radio room; a match cannot be resolved. Tuners are a benefit when used wisely. Tuners differ in their capabilities and circuits too. In general a modern tuner of proper design for solid-state output stages does better than older designs.

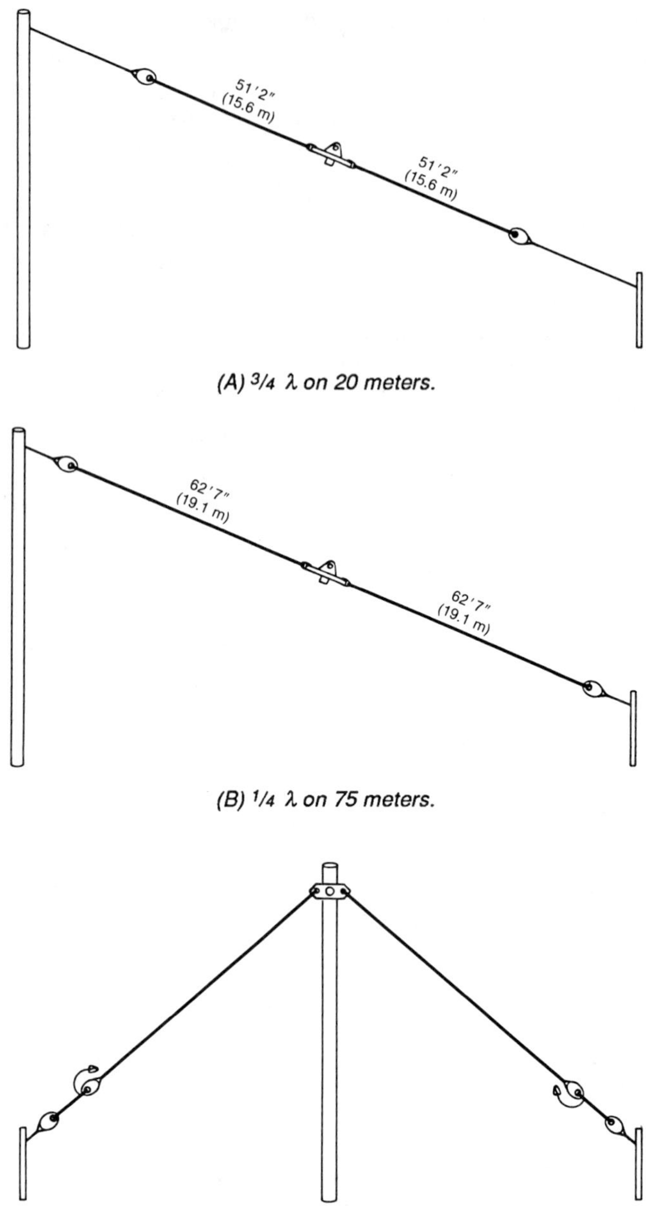

(A) 3/4 λ on 20 meters.

(B) 1/4 λ on 75 meters.

(C) Inverted-V plan.

Fig. 14-12. Two-band operation of sloper or inverted-V.

14-13. A 40-Meter Two-Element Triangle Antenna

The basic concepts and simple low-cost construction of the triangle antenna were presented in Topics 7-3 and 7-4. The triangle is well adapted to the low-frequency bands: 30, 40, 75, and 160 meters. On the 160-meter band there is a space problem because it is a full-wave loop. As mentioned previously, the triangle is not a true delta loop because it need not have equal-length sides. For low-frequency use, especially on 75 and 160 meters, the base must be longer than either side because of the height limitations.

A single loop for 80 and 160 meters has a feed point that is low and provides a match to 50–70 ohms coax lines. In like fashion, so does a low 30- or 40-meter single-loop triangle. As the height is increased the feedpoint resistance of 30- and 40-meter loops rises and a proper match can be made by using a 4-to-1 balun.

When a triangle reflector is used the antenna resistance drops and a direct match is made to coaxial line. A driven element and reflector combination requires only 0.12-λ separation and forms a space-saving pair.

Typically, the overall length of a 160-, 80-, or 40-meter triangle is calculated as follows:

$$\text{Triangle Wire Length, Driven Element in Feet} = \frac{984}{f_{\text{MHz}}}$$

Often the length must be trimmed a bit to obtain resonance on the desired frequency.

If the height of a 40-meter driven triangle is raised, its impedance rises. It may also be necessary to add wire to the total length to obtain resonance on the desired frequency.

The total wire length of the reflector is made about 5% longer:

$$\text{Triangle Reflector Length in Feet} = \frac{1030}{f_{\text{MHz}}}$$

The spacing equation is:

$$\text{Driven/Reflector Spacing in Feet} = \frac{118}{f_{\text{MHz}}}$$

A 40-meter driven triangle and reflector (Fig. 14-13) was constructed using much the same construction practice detailed for the 31-meter shortwave broadcast triangle of Topic 7-4. A 20-ft (6.09 m) length of 1.5-in (3.81 cm) ID PVC piping supports each of the two elements. Pieces of rope, stretched out each triangle base corner to a 5-ft (1.5 m) metal fence post. A pair of rope guys held each mast rigid.

First the driven element was erected and cut to resonance. The antenna impedance measured about 250 ohms and a 4-to-1 balun provided a good match to 50 ohms. The radiation pattern was bidirectional.

Next, the reflector was constructed. After it was placed in position the antenna resistance fell to about

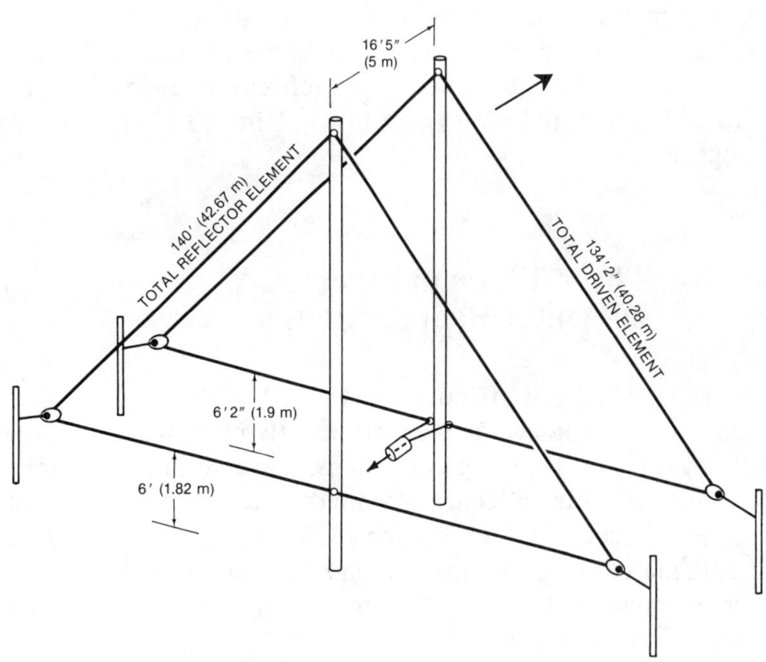

Fig. 14-13. Two-element 40-meter triangle.

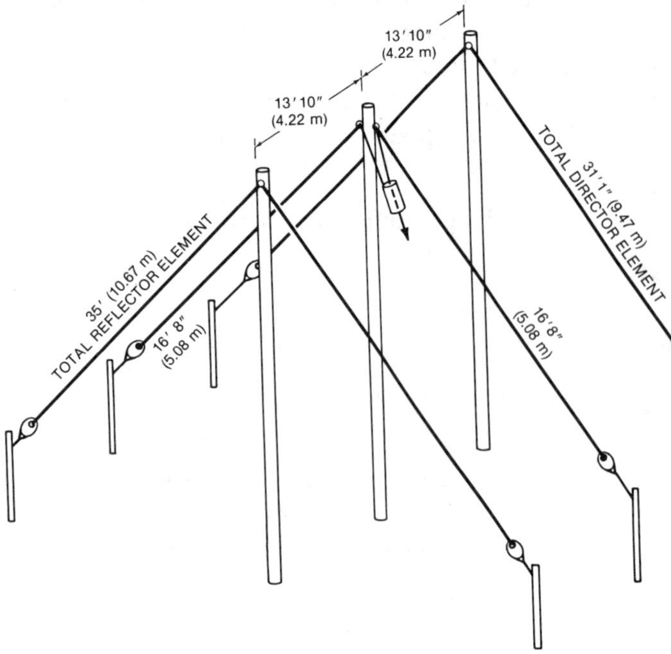

Fig. 14-14. A 20-meter three-element inverted-V beam.

50 ohms. However to reach 7200 kHz resonance it was necessary to cut off 1 ft 3 in (38 cm) from each end of the driven triangle. Remember to cut the same amount off each side of the antenna terminals because of the manner in which the triangle is formed from two like lengths of wire which are fastened together at the top of the mast. Also 1 ft 6 in (46 cm) cut backs were made on each side of the reflector where the ends join at the base of the reflector triangle. For 40-meter operation centered on 7.2 MHz the calculated values for total length of driven and reflector triangles are 136 ft 8 in (41.66 m) and 143 ft 1 in (43.6 m), respectively.

14-14. Three-Element Inverted-V 20-Meter Beam Antenna

An economical and effective fixed wire beam can be constructed using inverted-V elements. By using wide (0.2 wavelength) spacing between the driven element and the parasitic reflector and director, the antenna impedance does not drop to such a low value that difficult matching problems are encountered. In the arrangement of Fig. 14-14, a reasonable match is made to the 50-ohm line.

Start out with the dimensions given in Table 16-1 for driven element, reflector, director, and 0.2 wavelength spacing for a frequency of 14240 kHz. Some adjustments in length may be needed to obtain resonance, and good gain. Stretch the elements out horizontally as much as space permits. Note the reflector and driven element in Fig. 14-14 are a bit longer than table values. Use an antenna bridge to check resonance and impedance.

Received signal comparisons and some far away signal reports measured against a 20-meter ground plane vertical can help you determine gain and performance of this antenna. Switching between a beam and a ground plane as you receive signals from various directions can help you to evaluate front-to-back and front-to-side results.

My beam is angled at about 50° and provides peak results to Europe and the Near East with some good reports from the North Orient, Far East, and North Africa. If I erected a second fixed beam here I would obtain good results into much of South America and Australia with the beam peaked on New Zealand. In fact, in many parts of the world, two low-cost fixed wire beams will handle much of the earth's land mass at much less cost than the expensive beam and tower we often come to think is so very necessary. There can be a point at times when ham radio can become an obsession rather than fun and educational. Give this simple beam a try.

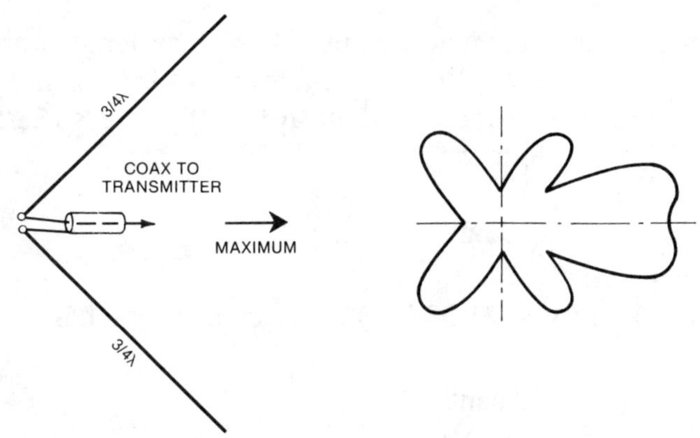

(A) Basic antenna and directivity pattern.

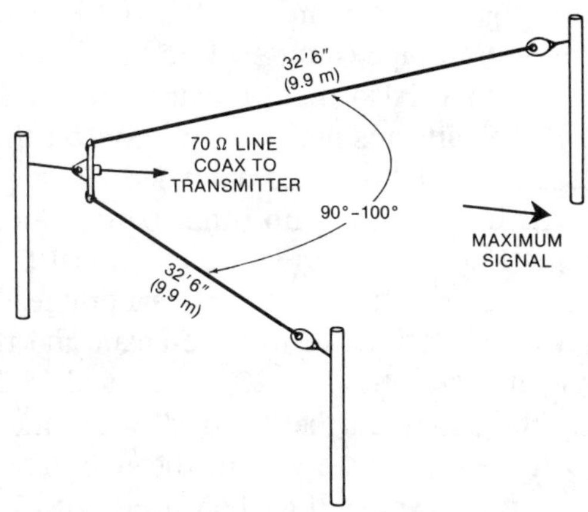

(B) A vee-beam antenna for 15 meters.

Fig. 14-15. The basic vee-beam and a practical example.

14-15. The Vee-Beam Antenna

The vee-beam antenna consists of two antenna wires tilted forward at an appropriate angle. Such an antenna displays directivity along a line that bisects the angle of the two wires as shown in Fig. 14-15A. The simplest and shortest vee-beam has 3/4-λ legs and just limited forward gain. The pattern shows that it has lobes in other directions also but is reasonably omnidirectional with a bit of gain in one direction.

The dimensions in Fig. 14-15B are for 3/4-λ operation on 15 meters (Table 14-1). However a tuner will also permit you to operate the antenna on 40 meters with a good match and an almost omnidirectional radiation pattern. A tuner will also permit a match on some other bands which will have to be determined by experimentation. Vee beams can be sloped forward as in Fig. 14-16 to improve bandwidth and directivity of low-height antennas.

Proper choice of the apex angle between wires directs the beam in a line that bisects the angle and delivers maximum gain in that direction. Table 14-3 is of special help in selecting the proper angle relative to length of antenna wires in terms of odd-multiples of one-quarter wavelength. The second row of the table gives the constants needed to calculate wire lengths for various odd multiples in feet; the third column states the angle between wires. For example a 3/4-λ wire length (in feet) on 21.3 MHz would be:

Table 14-3. Data for Planning Vee-Beams and Rhombics

Leg Wavelength	Calculation Constant	Vee-Beam Angles	Vee-Beam Gain (dB)	Rhombic Angles	Rhombic Gain (dB)
3/4 λ	726	100°	2.5	—	—
5/4 λ	1218	86°	3.3	—	—
7/4 λ	1710	76°	4.0	—	—
9/4 λ	2202	67°	4.75	104°	5.2
11/4 λ	2694	60°	5.3	94°	6.5
13/4 λ	3186	56°	5.8	86°	7.2
15/4 λ	3678	52°	6.3	80°	7.8
17/4 λ	4170	48°	6.8	74°	8.3
19/4 λ	4662	46°	7.2	70°	8.8
21/4 λ	5154	44°	7.6	67°	9.2
23/4 λ	5646	42°	8.0	64°	9.6
25/4 λ	6138	40°	8.4	61°	10.0
27/4 λ	6630	38°	8.8	58°	10.3
29/4 λ	7122	37°	9.2	56°	10.6
31/4 λ	7638	36°	9.6	54°	10.9
33/4 λ	8106	35°	10.0	52°	11.2
35/4 λ	8598	34°	10.3	50°	11.4
37/4 λ	9090	33°	10.5	49°	11.6
39/4 λ	9582	32°	10.7	48°	11.8
41/4 λ	10074	31°	10.9	47°	12.0

$$^{3}/_{4}\,\lambda = \frac{\text{constant}}{f_{\text{MHz}}} = \frac{726}{21.3} = 34 \text{ Feet}$$

The angle should be 100°.

The same arrangement with leg lengths of 81 ft 5 in and an angle of 86° operated as a 5/4-wavelength vee-beam on 20 meters (14.2 MHz). The calculation was:

$$^{5}/_{4}\,\lambda = \frac{\text{constant}}{f_{\text{MHz}}} = \frac{1218}{14.2} = 85 \text{ ft } 9 \text{ in}$$

However, the elements required some trimming to obtain exact resonance at the design frequency. An antenna bridge is a help.

Vee-beams can be tilted in the forward direction as in Fig. 14-16. By so doing there is an improvement in low angle radiation and the ends are readily accessible for trimming. Of course if you use a tuner, trimming is not necessary and you can tune up the antenna on other bands that provide proper loading. An 81 ft 5 in leg, 20-meter vee-beam loads and works well on 15 and 10 meters.

Table 14-3 provides data for both vee-beam and rhombic antennas (rhombics are covered in Topic 14-16). An example of a long vee-beam for 10 and 15 meters is illustrated in Fig. 14-17. The lengths are 39/4-λ wavelength on 10 meters and 29/4 λ on 15 meters.The calculated length for 10 meters (28.5 MHz) is as follows:

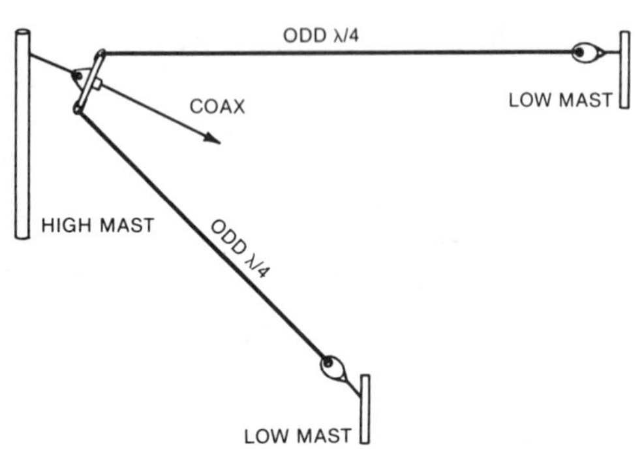

Fig. 14-16. Tilted vee-beam antenna.

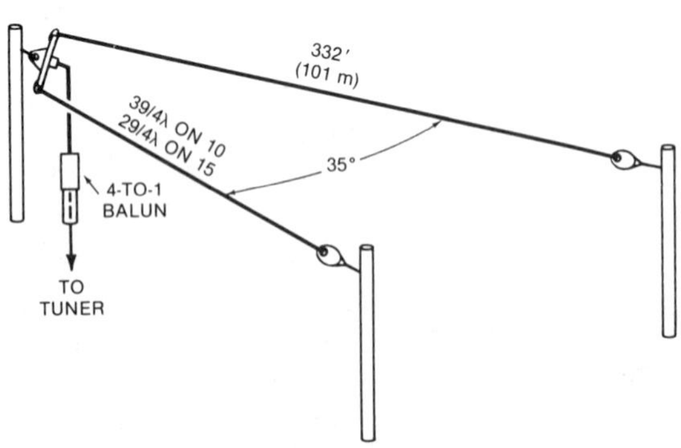

Fig. 14-17. Two-band long vee-beam antenna.

$$\text{Length} = \frac{9582}{28.5} = 336 \text{ ft}$$

For 15 meters (21.3 MHz) the calculated length is:

$$\text{Length} = \frac{7122}{21.3} = 334 \text{ ft}$$

Cut the length back to about 332 ft and use a tuner for a good vee-beam on 10 and 15 meters. Use an intermediate apex angle of 35° and you have a good performer. Also the 20-meter 19/4 -λ calculates out to 328 ft and results will be good on that band despite the lower than recommended 46° angle. The tuner will help loading on other bands, too. If the antenna is low you may require additional trimming. You can check results with an antenna bridge.

The long designs of the vee-beam and rhombic antennas of Figs. 14-17, 14-20, 14-21, and 14-23 were built by the author in the "sixties" when there were unused lots next to me and in the rear. They were great antennas especially when used with low-loss open-wire line and a tuner. An occasional letter or phone call about them comes along after 25 years. Try them if you have the space.

Today space limitations hamper most radio amateurs when it comes to the use of longwire beams. However you can fit a small vee-beam into a modest size lot and pick up some gain and bidirectivity on the higher frequency bands such as 10, 12, 15, and 20 meters. A good tuner will also load the beam on 30, 40, 80, and 160 meters although you must not expect too much in the way of gain. Try to raise the longwire antennas to as near the 30 ft (9.1-m) level as you can.

The examples of Fig. 14-18 loaded on all the previously listed bands using the MFJ Enterprises 941D tuner. Though not a super gain antenna you can achieve good results on all bands with one antenna. Typical tuner records for both lengths are given in Table 14-4. The table shows the proper settings of the three tuner controls for the sideband segments of the six bands of the first column. Such tables always help you to keep records so you can make band changes quickly for low SWR.

Usually the best results occur when the leg lengths are some odd multiple of a quarter wavelength

Table 14-4. Typical Tuner Records

Band	TX Match	Inductor SW	Ant. Match
61-ft Vee Beam			
160	4 1/2+	L	6
75	3 1/4	I	3 3/4
40	4 1/4	G	3
20	6	D	2 1/4
15	5 1/2+	B	2
10	5 3/4	B	1 3/4
126-ft Vee Beam			
160	6	L	4
75	2	I	4 1/3
40	4 3/4	G	3 1/2
20	5 3/4	D	2 1/2
15	4	B	2
10	2 3/4	B	1 3/4

or close to that figure. Note from Table 14-1 that a dipole cut for 75 meters or 160 meters has a number of harmonic lengths on the higher-frequency bands. For example, a 126-length dipole for 160 meters also is near to odd multiple lengths on 10, 12, 15, 20, and 30 meters. A tuner can reduce the SWR to a low value and the antenna will perform well on these bands. A 75-meter dipole cut to 61 ft 6 in has near-resonance on 10 and 15 meters and will also load on all of the other bands. In fact the MFJ Enterprises 941D will even load the 75-meter 61 ft 6 in length on 160 meters. An appropriate compromise angle between the lengths can be determined from Table 14-3 depending upon the available space of your installation site. These antennas will certainly not be dynamite super vee-beam antennas but they will give you good results on all bands with the use of a tuner.

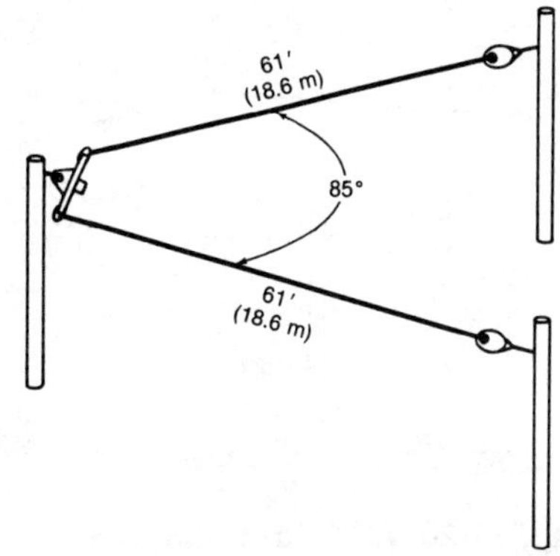

(A) 61-ft vee beam.

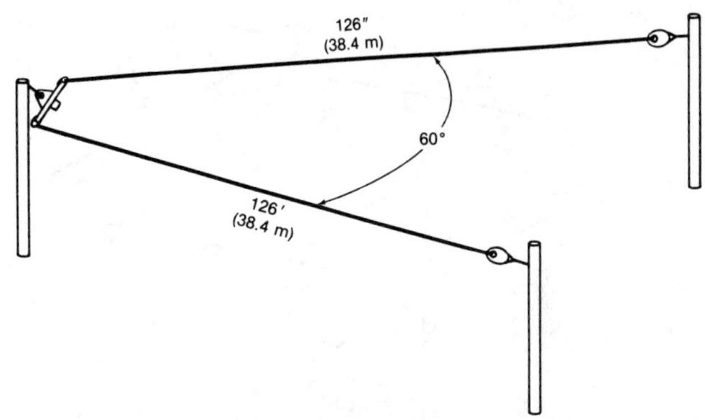

(B) 126-ft vee beam.

Fig. 14-18. Short vee-beam antenna for use with a tuner.

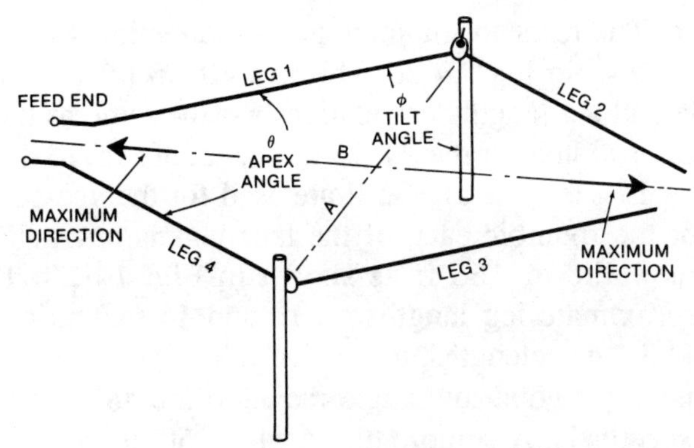

Fig. 14-19. Resonant rhombic antenna.

14-16. Rhombic Antennas

The rhombic is a diamond-shaped longwire antenna (Fig. 14-19). For a given length of wire it has a somewhat higher gain than a vee-beam but the space requirements are longer and narrower. Additional support masts are also needed.

Also important is the *wave angle* or vertical angle of radiation. For long distance the wave angle should fall between 10° and 20°. A rhombic elevated a wavelength or more above ground takes maximum advantage of a low wave angle. Rhombics like vee-beams can be tilted forward to favor the low angles if you can't take advantage of great height. For example in Fig. 14-20 masts 2 and 3 can be somewhat lower than mast 1 while mast 4 can be lower than masts 2 and 3.

In a resonant rhombic the two legs on each side (L1 + L2) and (L3 + L4) should be an odd multiple of a half wavelength. Angles and lengths can be derived from Table 14-1 and the rhombic information given in Table 14-3 of Topic 14-15.

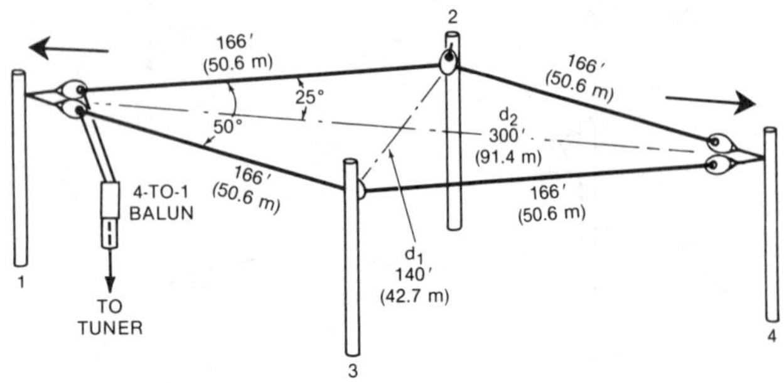

Fig. 14-20. A 10-15 rhombic with balun.

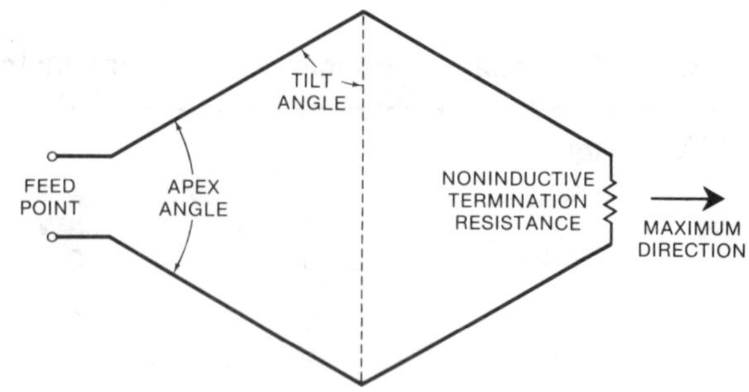

Fig. 14-22. Basic terminated rhombic.

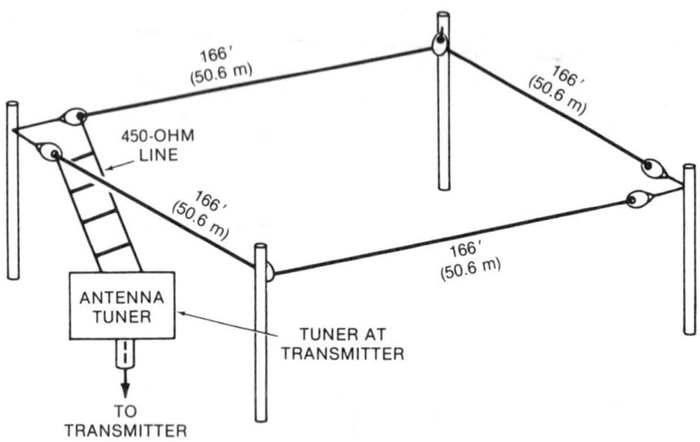

Fig. 14-21. Rhombic with open-wire line and antenna tuner.

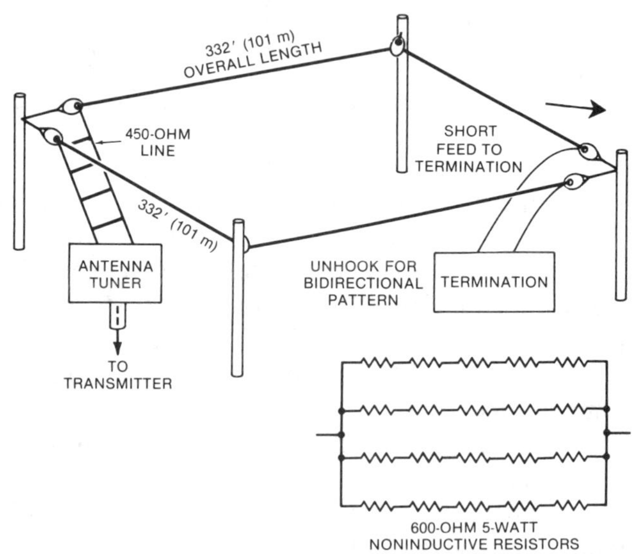

Fig. 14-23. Practical terminated rhombic antenna.

The resonant rhombic has a bidirectional pattern as shown in Fig. 14-20. This is a 10–15 meter rhombic and leg lengths calculations are the same as those demonstrated for the 10–15 vee-beam of Fig. 14-17.

Leg lengths of 332 ft are used for the vee-beam. For the rhombic each of the four legs is one half of this value or 166 ft as shown in Fig. 14-20. The approximate leg lengths on 10 and 15 meters are 5 and 3.5 wavelengths, respectively. The preferred apex angles for good low angle radiation are 48° and 56°, respectively. A compromise angle of 50° is satisfactory. This rhombic loaded properly and performed well on 20 meters also. I prefer the use of open-wire line as shown in Fig. 14-21, if the tuner has a balanced input. I located the tuner at the transmitter and the overall length of open-wire line is about 150 ft. Very little line loss is involved.

Simple trigonometry can be used to locate the four support poles and set up the correct apex angle. The distance between side poles is:

$$d_1 = 2(\sin 25^\circ \times 166) = 140 \text{ ft}$$

The distance between the near and end poles is:

$$d_2 = (\cos 25^\circ \times 166) = 300 \text{ ft}$$

The resonant rhombic is a bidirectional antenna. However the rhombic can be made unidirectional by employing a resistive termination at the far end of the antenna as shown in Fig. 14-22. The characteristic impedance of the rhombic configuration approximates 800 ohms. An 800-ohm noninductive resistive termination at the far end makes the rhombic nonresonant and unidirectional.

When the rhombic is terminated properly, the antenna resistance at the feed point is 800 ohms. An adequate match can be made to 450-ohm parallel wire line. Use an antenna tuner with a balanced input (Fig.

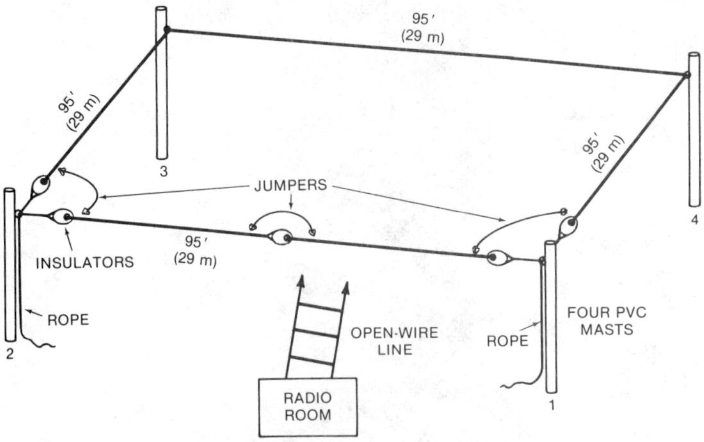

Fig. 14-24. A small-space rhombic antenna, with two bidirectional and one omnidirectional connections.

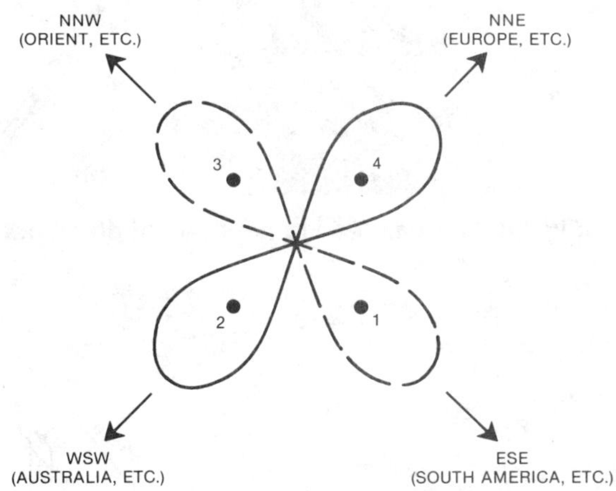

Fig. 14-25. Bidirectional patterns of the antenna in Fig. 14-24.

14-23) and excellent results are obtained with low back pickup.

The network used to terminate the rhombic antenna consists of 20 five-watt, 600-ohm noninductive resistors connected in a series-parallel network with a net resistance of 750 ohm. Such a network is capable of handling 200 watts PEP. You will have to do some searching to find noninductive resistors.

Do you have a plot of ground approximately 100-feet square? If so, an antenna mast mounted at each corner will support the versatile and good performing pseudorhombic of Fig. 14-24. The angle is not ideal and leg lengths are not very long but the antenna does work as well as many expensive beams and it has multiband capability. The performance is excellent on 10, 15, and 40 meters. It also loads and produces acceptable results on 80 and 160 meters, but does not perform quite as good on 20 meters.

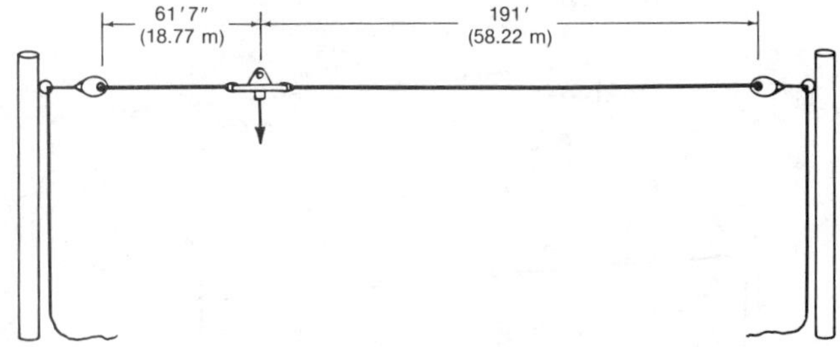

Fig. 14-26. Full wavelength antenna for the 75-meter band.

Mount the small-space rhombic close to the radio room if you can and run a short length of 450-ohm open wire line to it. This arrangement gives you the means of changing to the two bidirectional patterns of Fig. 14-25. This is accomplished by moving the open-wire line to pole 1 or pole 2. Lanyards are attached to both poles so the wires can be lowered readily for making feed and operational changes. For omnidirectional operation you can jump both corners and feed the center of the leg between poles 1 and 2. When you feed at pole 1 be sure to connect the jumpers across the center insulator and the ones at pole 2. When you feed pole 2 connect the jumper across the center insulator and the one at pole 1.

14-17. Full Wavelength Antennas

The full wavelength antenna comes in many forms—from a linear straight wire, to a quad, or a delta loop. Actually, the folded dipole is really a folded full wavelength of wire or tubing. The easy-to-build triangle covered in this book is a full wavelength style. The quad is a popular and top-performing full wavelength antenna. When you wish to become active with the more involved construction of quads pick up a copy of Bill Orr's (W6SAI) book on Cubical Quad Antennas*. In this topic an easy-up horizontal quad is presented along with some unusual variations. A linear full-wave antenna is also covered along with some other types of loops.

Modern transmission lines and transmitters operate at low impedance. A typical way of feeding a full-wave linear wire is shown in Fig. 14-26. The feed

*Radio Publications, Inc. (See Appendix)

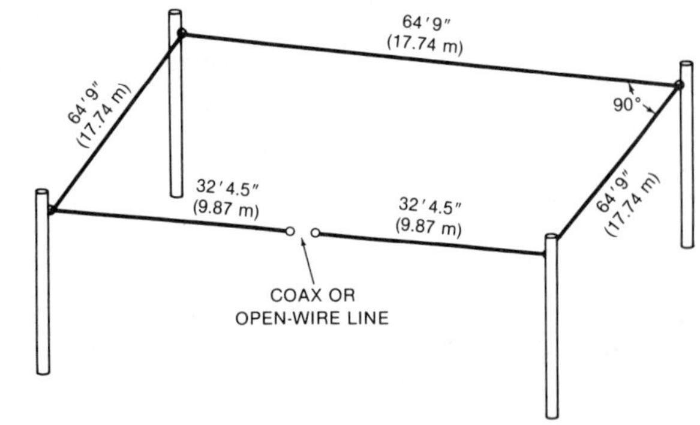

Fig. 14-27. Horizontal quad loop for 75 meters.

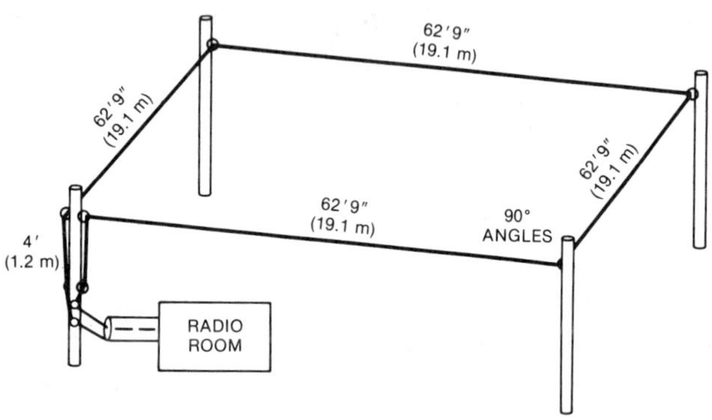

Fig. 14-28. Diamond fed quad.

point is a quarter wave from one end of the antenna and three quarter waves from the other end. This reflects a low impedance at the feed point and provides a reasonable match to a coaxial line.

The antenna shown in Fig. 14-26 is dimensioned for 75-meter operation using the lengths given in the Chapter 16 tables (1/4 and 3/4 columns). The resonant frequency is about 3.725 MHz. Cutting off 1 ft 6 in from the short end and 6 ft from the long end moves the resonant frequency up to 3.8 MHz.

The bandwidth of the antenna is rather narrow and for operation over the entire band a tuner is suggested. Such an antenna is a single-band type and it does very well on that band. A tuner can load it on other bands but performance is erratic, depending upon frequency.

14-17-1. The Basic Horizontal Loop

The horizontal single-band full-wave loop of Fig. 14-27 is an attractive antenna for stretching around the borders of a square or nearly square lot. It does a fine

Fig. 14-29. Coax feed to near end of diamond.

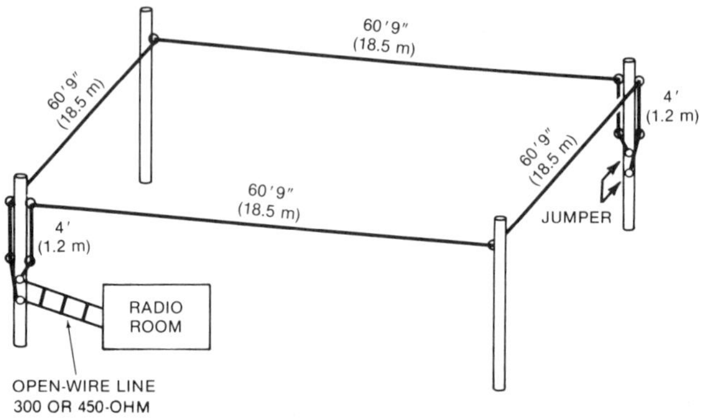

Fig. 14-30. W3FQJ diamond installation.

job even at low mounting height. If you can't stretch out a full-wavelength linear antenna such as Fig. 14-26 try a loop. Loops are popular when you are surrounded by trees. Although they may wind and dip among the trees, they work out well.

Good results are obtained on 40, 75, and 160 meters with a loop antenna. Cut the loop about 2 percent longer than the full-wave free-space dimension given in the table of Chapter 16. It is usually easier to cut back on the length of the antenna than add length to reach a specific resonant frequency.

If trimmed exactly you can usually connect it

Fig. 14-31. Far end mast base showing jumper for closed or open operation of diamond loop.

directly to a coaxial line and obtain an SWR below 1.5 to 1. A tuner can be used for exact matching. A tuner also permits operation on other bands. My own experience has not been rewarding whenever I have operated closed loops on any band other than for which it has been cut when compared with other types of antennas at the same height. Even a single vertical quad does not do well when tuner-loaded on another band.

14-17-2. The Diamond Loop Antenna

In my own situation with many trees and other antennas it was convenient to try out an equal sided diamond with 90° corners (Fig. 14-28). The diamond is fed at one of its corners which is brought near the radio room window. The results are good despite low antenna height and considering it was guided through trees and around obstacles. The three poles are 19 ft (5.8 m) high and the fourth one only 10 ft (3.05 m) at

Table 14-5. Typical Tuner Settings for Diamond Loop

Meter Band	Match	Inductor	Antenna	kHz
Diamond Closed				
75	3	L	$3\frac{7}{8}$	3800
Diamond Open				
160	4	L	$5\frac{7}{8}$	1850
75	3	I	6	3800
40	$2\frac{1}{8}$	G	$5\frac{7}{8}$	7200
20	2	D	$1\frac{7}{8}$	14200
15	$3\frac{1}{3}$	C	$5\frac{7}{8}$	21300
10	3	B	6	28500

the feed corner near the radio room. With the lengths shown in Fig. 14-28 the resonant point is located near 3820 kHz. A short 20-ft (6.1-m) length of coaxial line connected the loop to the transmitter with a good match (Fig. 14-29). A tuner permits low SWR operation over the entire band.

14-17-3. All-Band Diamond Antenna

My final version of the diamond antenna is shown in Fig. 14-30. In this arrangement the far end of the diamond is brought down to a low level (Fig. 14-31) where it can be made accessible for operation as either a closed or an open diamond. Better operation on 75 meters was obtained with the end closed. However on other bands better results were possible with the diamond end open. A short jumper accessible from ground level permitted an easy changeover.

When the short jumper was open the diamond tuned up on 75 meters as well. Although it was not quite as good as the closed connection, it provided a true all-band antenna with good performance. When you want particularly good operation on 75 meters you can always connect the jumper into position. Be certain to keep a record of tuner settings as shown in Table 14-5. The ones shown are for my installation and the MFJ Enterprises MFJ-941D tuner.

In tuning up longwire antennas of this type you can often obtain a low SWR on more than one setting of the tuner. Many modern tuners allow you to choose the inductor switch setting that provides maximum output.

Two guy ropes support (Fig. 14-32) each corner mast. At the feed end of the diamond the antenna wires are dropped down the mast to two accessible

Fig. 14-32. Top of mast showing right-angle turn of antenna and the support guys.

Fig. 14-33. Open-wire feed of diamond.

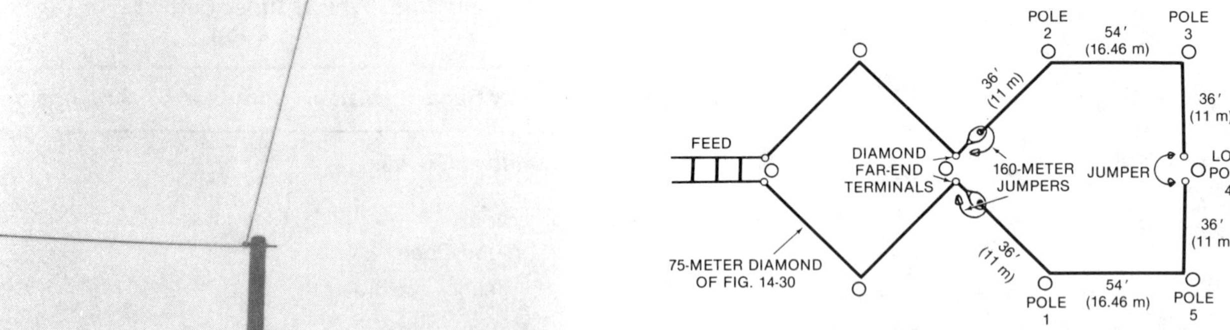

Fig. 14-34. A 160-meter hybrid loop layout.

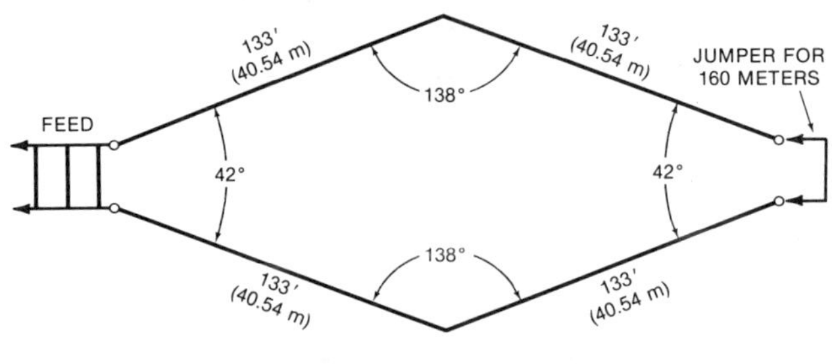

Fig. 14-35. Bidirectional rhombic and 160-meter closed loop when far end is shorted.

antenna terminals (Fig. 14-33) where a short length of open-wire line connects the diamond antenna to the balanced input of the MFJ Enterprises MFJ-941D tuner which includes an internal 4-to-1 balun. Fine results were obtained on all bands except 20 meters where results were comparable to a 20-meter plane vertical with its top at the same height.

14-17-4. 160-Meter Addition to 75-Meter Diamond Antenna

The arrangement of Fig. 14-34 shows a method of adding 160-meter full-wave capability to the 75-meter diamond discussed in the last section. Five additional PVC masts are needed as indicated, forming a rectangular full-wave loop with an indentation. The 160-meter loop can be placed in operation by placing jumpers across the two insulators which connect to the two end terminals of the 75-meter diamond.

The total area occupied by the antenna is about 130 ft by 75 ft (39.6 m by 22.9 m) and provides all-band 10–160-meter performance using a tuner.

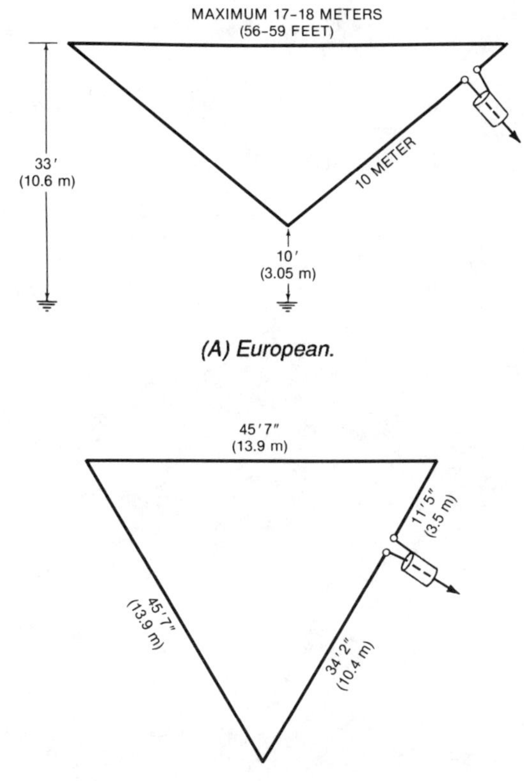

(A) European.

(B) American.

Fig. 14-36. Special European delta loop.

Operation on 160 meters is better with the far end closed than open. On other bands it works better with the far end open. If ample space is available you can stretch out a 160-meter end-fed loop or diamond, especially if 160 meters is a special band for you. Use the arrangement of Fig. 14-30 with 129-ft sides and an open-wire line from terminals to tuner.

Another approach is to use a rhombic antenna. Despite the narrower feed and far-end angles (Fig. 14-35) you can obtain loop-like results on 160 meters if each leg of the rhombic is made 133-ft (40.54-m) long (total length of 532 ft or 162.2 m). You will need a lanyard to pull down the far end of the rhombic to connect the short for 160-meter operation or add a terminating resistor to make your rhombic unidirectional.

The (L1 + L2) and (L3 + L4) sides of the rhombic have a 266-ft length which falls within tuning reach of odd-multiples of a quarter wavelength on 30, 20, 15, 12, and 10 meters when you use open-wire line and a tuner with a balanced input. On 21.3 MHz, $^{23}/_{4}$ wavelength is almost exactly 266 ft. Use an angle of about 42°. Twenty meters is usable with 259 ft at $^{15}/_{4}$ wavelength.

14-17-5. Special Delta Loop

The delta loop is a common full wavelength antenna in Europe and is operated on the lower frequency bands to a great extent. John I2POR (Italy) uses the specific configuration shown in Fig. 14-36A on 40 meters. He has worked 45 countries and attests to its low wave angle and vertical polarization that permits good DXing and a cut back in the level of closer-in signals. This characteristic is aided when the loop is fed one quarter wavelength from the bottom corner.

John's own antenna has its horizontal side approximately 33 ft (10.6 m) above ground while the bottom corner is about 10 ft (3.05 m) above ground. Such an antenna has a limited bandwidth and must be trimmed carefully to a preferred 100-kHz segment of the band.

A true delta version for operation centered about 7200 kHz here in the USA would have the dimensions shown in Fig. 14-36B. However, as shown in the basic arrangement of Fig. 14-36A, the top length can be somewhat longer than the sides if need be to better accommodate the height limitations at your erection site. A tuner is helpful in obtaining a good SWR when you wish to operate on other portions of the band with only a limited deterioration in antenna performance.

Chapter 15

Antennas for the New Technician and New VHF/UHF Operator

THE antenna descriptions in this chapter are directed mainly to the new amateur radio technician. However much of the material is also appropriate for higher grade licensees who are about to become active on the VHF/UHF bands. Most new technicians begin their initial operation on 2 meters and antennas for this band are emphasized. The 2 meter/70 centimeter duoband transceiver is attractive to the new and the more experienced technicians. Again it is important to remind you that addresses of manufacturers and other information sources are given in the Appendix as well as the end of Chapter 1.

15-1. Introduction

At present there are no significant FCC changes contemplated for the 6-meter band. Operation on this band will remain exciting with the occasional opportunities for some long distance DXing. Six meters with its various propagation modes can provide some unusual contacts. Antennas can be varied and interesting. This band represents an ideal place for the technician to learn much about easy-to-erect and low-cost antennas of various configurations.

The wire type antennas described in previous chapters will work well on this band. Use the table in Chapter 16 for correct lengths. Information and construction guidance for verticals, dipoles, and simple beams are detailed. The Chapter 16 dimension tables present typical lengths and spacings in inches for the various VHF/UHF bands. Although the antennas discussed may be based on lengths for a specific band you can use the tables for dimensioning the same type of antenna for operation on another band. Several multiband antennas are described in this chapter.

Data is given on how to check and compare VHF/UHF antennas. The importance of reference antennas, check frequencies, and SWR measurements are important subjects of this chapter.

15-2. VHF/UHF Quarter-Wave Vertical Antennas

The quarter-wave vertical is the most common mobile VHF/UHF antenna in use. The car roof serves as an effective ground plane. An Antenna Specialists quarter-wave mobile antenna with a magnetic base is shown in Fig. 15-1. I often use one atop a metal rack in the radio room. A magnetic base-mount antenna is an attractive way to go for the new technician. The antenna can be moved about easily and hidden away in the car when you are parked. The photograph shows the coaxial line as it is fed into the car through the rear door. You may also wish to use an easy-removal mount for your transceiver so you can use it both in the car and in the house as well. This is an economical way of starting your technician activity with a single unit and a simple antenna.

Fig. 15-1. Antenna Specialists 2-meter magnet-mount antenna centered on roof of station wagon.

Fig. 15-2. Antenna Specialists 2-meter ground-plane vertical.

The inexpensive ground-plane vertical of Fig. 15-2 can be your home antenna, completing your first versatile ham installation. The Antenna Specialists ground-plane vertical attaches very easily to a 1-in ID 10-ft top section of a PVC mast (Fig. 15-3), but it also can be held by a chimney strap, vent pipe clamp or a roof or sidewall mount. The ball at the top of the vertical has a set screw and can be moved up or down to permit you to set the minimum SWR to a desired spot in the 2-meter band.

The Celwave antenna arrangement of Fig. 15-4 is versatile because it can be used on more than one band. A single mount is attached to the car. Four separate stiff wire radiators can be cut for resonance on the 50-, 144-, 220-, and 420-MHz bands. The 2-meter and 70-centimeter versions are shown in Fig. 15-4. The 2-meter radiator of Fig. 15-4A is 19 in (48.3 cm) and the 70-cm radiator is 5.25 in (13.34 cm). When you change bands all you need do is insert the proper radiator into the mount. The graphs of Fig. 15-5 help you cut the whip to a specific frequency or band from 135 to 400 MHz. For frequencies between 425 and 450 MHz the recommended length is 5.25 in.

Fig. 15-3. A $\lambda/4$ ground-plane vertical atop mast.

(A) AM whip.

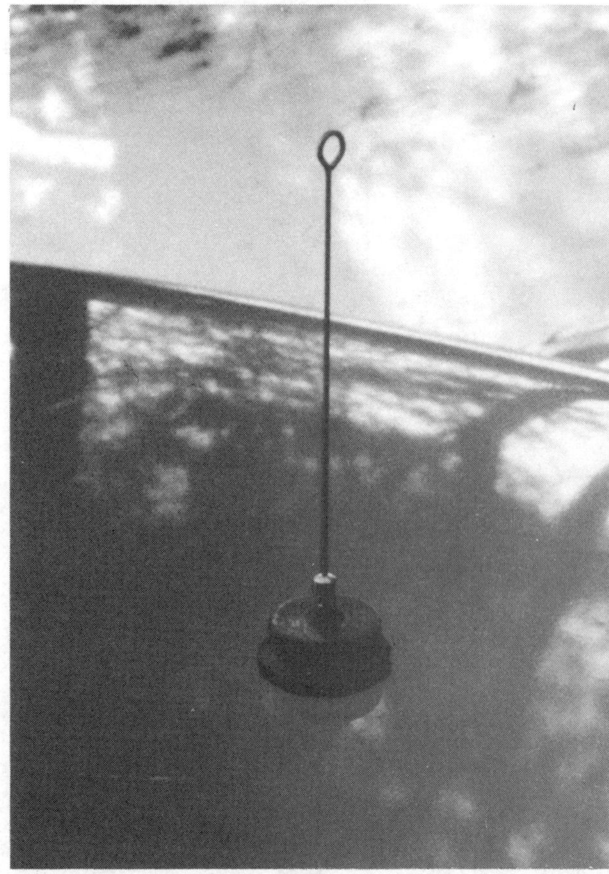

(B) 70-cm whip.

Fig. 15-4. Celwave vertical and exchangeable whips.

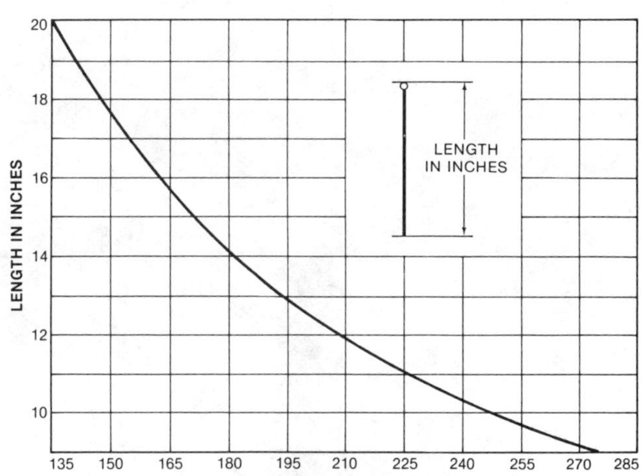

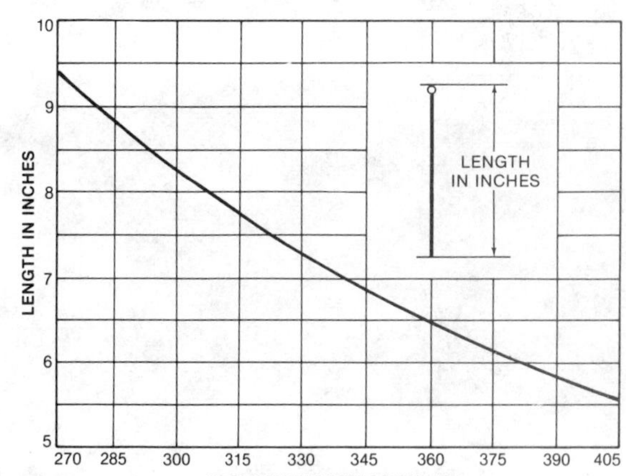

Fig. 15-5. Cutting lengths for whip. (Courtesy Celwave Systems, Inc.)

The mount is a low-profile one. The base has a diameter of only $1^{1}/_{2}$ in. A $^{3}/_{4}$" mounting hole is required. The coaxial cable is fed through the hole and finally the bushing to which the cable is attached is tilted and rocked through the hole. Center the bushing and tighten down the locking ring. The antenna base is screwed down over the locking ring. Only the cable must be guided to the transceiver. It is cut to the

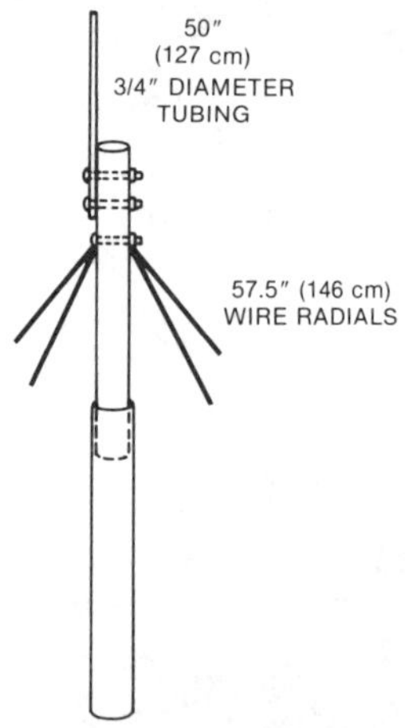

Fig. 15-6. A 6-meter elevated ground plane and PVC mast.

correct length and a PL-259 plug is attached. Note that whips can also be cut for landmobile, aviation, and other radio services.

A homebrew 6-meter vertical is shown in Fig. 15-6. The construction details were given previously in Topic 13-5. All you need to change is the length of the tubing to 50 inches (129.5 cm) (which allows 1-in for influence of bolt) for 6-meter operation and cut the radial wires a length of 57.5 in (146 cm).

An inexpensive CB antenna consisting of the Radio Shack luggage rack mount (21-937) and a CB whip (RS-21-903) can be used as a 6-meter vertical. Cut the whip down to 53.75 in (136.5 cm) for 6-meter operation. The mount and antenna can be used as a mobile installation (Fig. 15-7) or it can be bolted to the top of a PVC mast for a home installation. Appropriate 57.5-in radials must be attached as shown in Fig. 15-8. Here the radials are ready to stretch out.

15-3. Checking the General Performance of a VHF/UHF Antenna

Information about checking out ham antennas was given in Chapter 12. The importance of SWR measurements was also covered. Transmission line losses

Fig. 15-7. Luggage rack mount and 6-meter whip.

Fig. 15-8. Luggage rack mount bolted to top of PVC mast with 6-meter vertical whip.

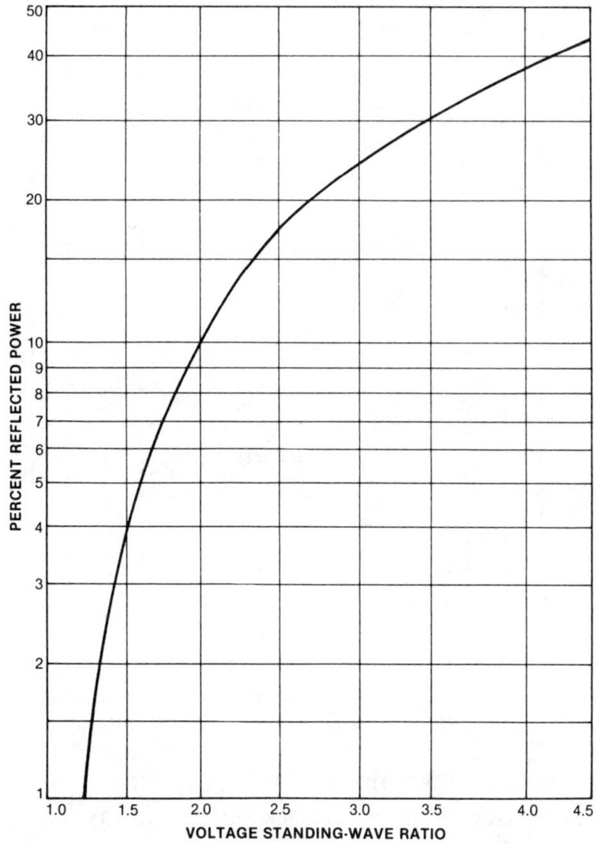

Fig. 15-9. Percent reflected power vs standing-wave ratio.

are higher in the VHF/UHF spectrum and it is important to use good quality coaxial line. The maintenance of a low SWR helps to minimize the overall line losses and also permits your transmitter to deliver its maximum output. Remember again that modern solid-state transmitters often initiate their own power output decline for protective reasons when they are mismatched. Thus a high SWR can give you two specific losses—line and transmitter. The useful curve of Fig. 15-9 shows the percentage power loss as a result of mismatch. However it does not show line attenuation which is related both to the line quality and the SWR. Also it does not show how much your transmitter is backing off as a result of the mismatch. All these stress the importance of maintaining a good match and low SWR.

Much about the performance of a new antenna can be evaluated on receive as also detailed in Chapter 12. For FM the same general techniques apply when receiving a simplex signal from a fixed station or a parked mobile. The procedure also applies to the reception of a distant repeater. Remember you have quieting in an FM receiver and all received signals above a certain level result in the same level output. Full quieting is reached and you cannot expect to notice any difference in performance between two antennas when receiving a strong local signal.

Much of my own antenna checking is on distant repeaters, switching between the antenna under test and a reference antenna as the received signal strength is displayed. Also a contact through one of the more distant repeaters can help with transmit comparisons as well and even via an occasional simplex schedule.

Choose your testing frequency carefully so you do not interfere with repeater operations. Choose frequencies designated for general simplex operations and make certain you do not cause interference to a contact under way. Pay attention to the instructions to be found in the ARRL Repeater Directory. In checking your SWR find suitable test frequencies at the center and two ends of the desired operating frequency range. By so doing you can make antenna adjustments that can center the minimum SWR in the range.

When the SWR is too high, check your installation carefully if you have purchased a commercial antenna. Go over the instructions supplied by the manufacturer to make certain they have been followed. If the antenna is a homebrew one, length adjustments may be necessary and, in the case of a directional antenna, spacing and/or the antenna matching system may require a change. Remember that at VHF/UHF frequencies a fraction of an inch can sometimes make a significant difference. It is best to get the SWR situation straightened out as best you can (better than 2 to 1) before making any extensive receive or on-the-air transmit tests.

15-4. A Two-Meter Beam Antenna

A small three-element beam contributes good gain and directivity. Aiming is not as critical as that required by a multielement beam and, although mounted in a fixed position, such as antenna displays good performance over a rather wide arc. The half-power angle (beamwidth) of the HyGain 23B-S Telex beam of Fig. 15-10 is 90° when mounted vertically and 60° when mounted horizontally. The gain is 8 dB

Fig. 15-10. HyGain 3-element 2-meter beam.

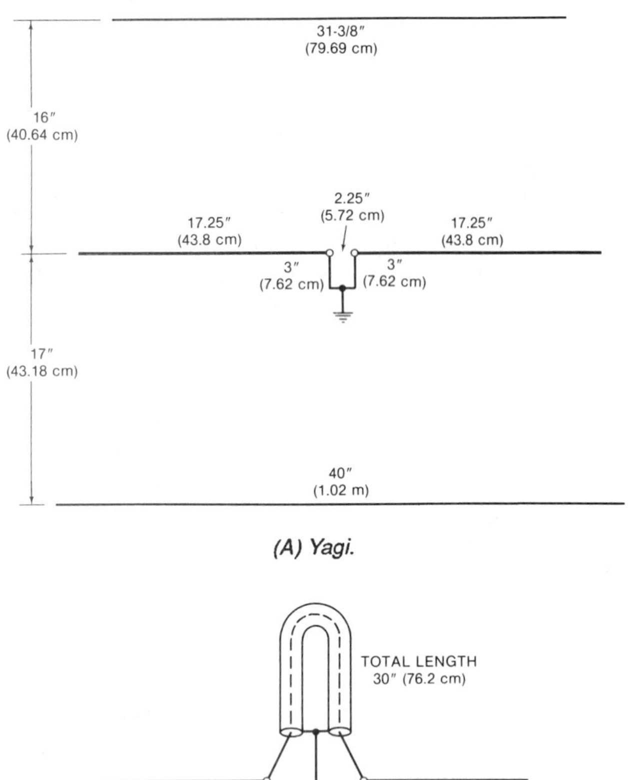

(A) Yagi.

(B) Balun.

Fig. 15-11. General dimensions of HyGain Yagi and the balance-to-unbalance 4-to-1 coaxial line balun.

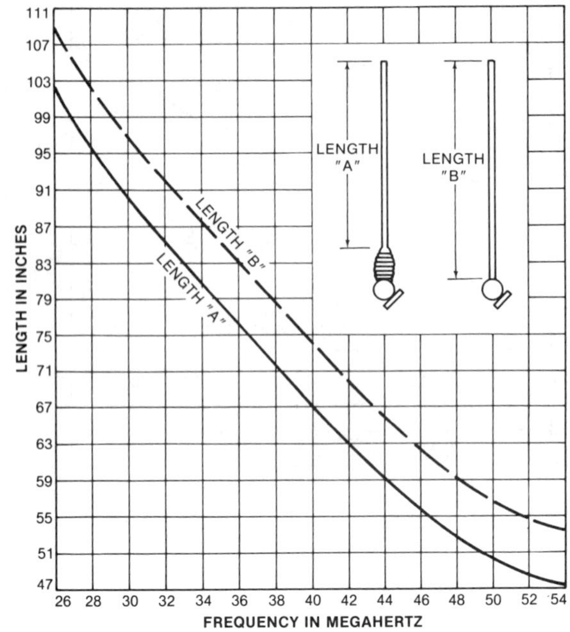

Fig. 15-12. Swivel ball mount and whip-cutting data. (Courtesy Celwave Systems, Inc.)

and front-to-back ratio is 20 dB. Thus there is significant rejection of side and back pickup. In my location when angled due west there is a good sweep of land accommodated in a favorite direction with a 90° beam width.

A combination of an omnidirectional vertical and a beam is an economical approach for many two meter operators. No rotator is necessary, only two lines and a two-position UHF coaxial switch are used. The beam can be aimed in some useful or favorite direction, or towards a favorite repeater some distance away. Often such a fixed beam is of help in making contact via simplex with a mobile that travels the same commuter path each day. This technique can be used without the time limitation of a busy repeater.

The general plan and dimensions of a HyGain Telex beam are given in Fig. 15-11. Note the small U-loop that extends away from the driven element and is grounded at its center to the antenna support boom. This is called a "stub" or "beta match" which makes certain the impedance across the terminals of the driven element is 200 ohms. A 4-to-1 balun made from a half-wavelength segment of transmission line transforms this impedance to 50 ohms to match the 50-ohm cable that runs to the transceiver. (The beta loop and the balun can be seen in Fig. 15-10.) The balun is one of several methods of matching directional anten-

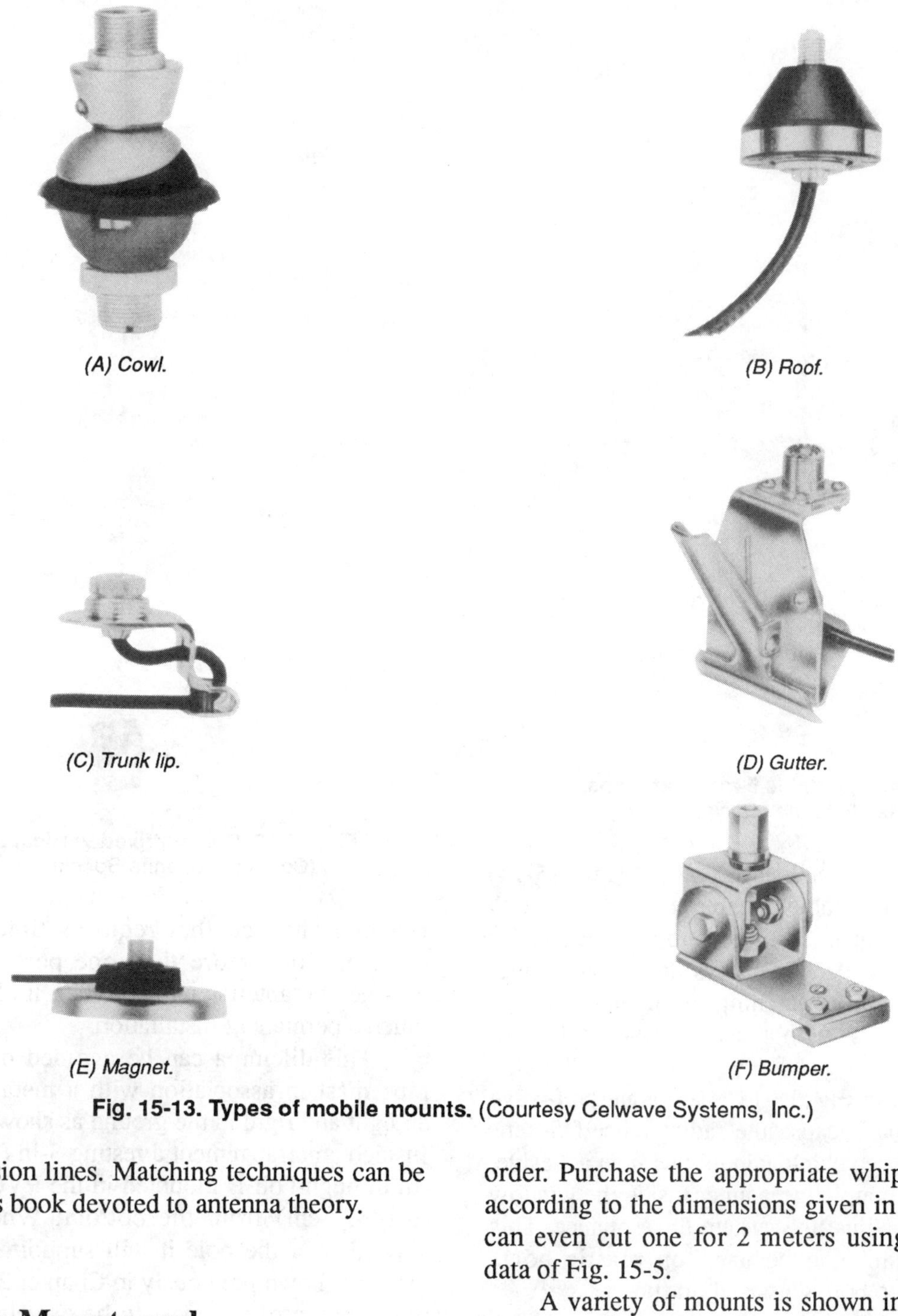

(A) Cowl. (B) Roof.

(C) Trunk lip. (D) Gutter.

(E) Magnet. (F) Bumper.

Fig. 15-13. Types of mobile mounts. (Courtesy Celwave Systems, Inc.)

nas to transmission lines. Matching techniques can be found in various book devoted to antenna theory.

15-5. Mobile Mounts and Mobile/Portable Antennas

Various mounts are available for attaching mobile antennas to a vehicle. The simple ball-swivel mount with or without a spring is a common mount for CB, 10 meters, and 6 meters. Such a mount is readily available at ham or two-way radio stores, Radio Shack and similar outlets, auto stores, and by mail order. Purchase the appropriate whips for it and cut according to the dimensions given in Fig. 15-12. You can even cut one for 2 meters using the dimension data of Fig. 15-5.

A variety of mounts is shown in Fig. 15-13. An advantage of the gutter and magnet mounts is that they can be removed easily and hidden in the car. If you plan to operate mobile on several bands it often pays to buy mounts and components separately and construct your own combination that can be interchanged conveniently when changing bands. Shop around the hamfests and flea markets. Often you may find bargain rates for broken or used antennas with good mounts.

Fig. 15-14. A portable 6-meter antenna.
(Courtesy Antenna Specialists, Inc.)

Fig. 15-15. Glass-affixed vertical antenna.
(Courtesy Antenna Specialists, Inc.)

The Antenna Specialists HM-2010 (Fig. 15-14) is an example of a portable 6-meter antenna. Although a fixed-mount antenna, it collapses to a small size for taking out on VHF/UHF field days. It is a half wave-length antenna with a matching arrangement at the base. The gain is 3 dB with a low-angle vertical radiation pattern.

The antenna in Fig. 15-15 is a unique half-wave 3-dB gain antenna because the radiator itself fastens to the outside of the vehicle window with an adhesive. It is capacitively coupled to a tuner inside the window which is used to adjust the system to resonance. This style antenna can also be used on plastic boats because it clings to both glass and plastic.

15-6. An Easy Arm-Powered Rotator

In addition to cost there are other disadvantages to a rotator that actually discourage experimentation and antenna system change. A rotator involves additional hardware of considerable expense and a strong rigid mount is needed that requires time and effort to change. Often more than one person is needed to replace one antenna with another. It all becomes very much a permanent installation.

This dilemma can be avoided by using a PVC pipe mast in association with a metal pole mounted straight and rigid in the ground as shown in Fig. 15-16. In such an arrangement a resting 3-in (7.6 cm) long 1/4 -in through-bolt is mounted in the PVC pipe about 18 in (45.7 cm) from the bottom. When the mast is slipped over the pole it will support the mast above ground (shown previously in Chapter 2). The pole permits the top of the mast to be elevated an additional few feet above ground. Furthermore, the mast provides enough rigidity to maintain any small lightweight VHF/UHF beam such as two or three elements on 6 meters, three or four elements on 2 meters, and seven or eight elements on 70 cm in a given direction. A bit of twisting motion permits you to rotate the beam in another direction when making checks and comparisons with an omnidirectional reference antenna.

Fig. 15-16. A 2-in ID mast base slipped over metal fence pole.

A firmer lock can be established by using an eye-ring tapered wood screw a bit larger in diameter than a hole drilled in PVC pipe as shown in Fig. 15-17. In the example shown the eye-ring screw is twisted into a 1/4-in hole drilled into the PVC pipe about 6 in below the resting bolt. When you screw this eye-ring into the hole its point soon presses tightly into the side of the metal pole, preventing the mast from shifting orientation in the wind. Loosen the eye ring up a bit and you can change antenna direction with ease. If you want an extremely tight bind you can pass a stout screwdriver through the eye and use it as a wrench. Three slack guys located near the mast top will prevent the mast from whipping too much in high wind. At the same time you can orient the beam through 360° with little difficulty.

Fig. 15-17. Close-up of mast showing resting bolt and tapered screw eye that pinches metal pole and holds mast in fixed position.

15-7. The 6-Meter and 70-Centimeter Beams

Beams are useful on 6 meters through 70 cm for DXing, contests, and other activities. Again, an omnidirectional antenna and a fixed directional one are a convenient operating pair adequate for many situations. If needed you can use the arm-power rotation arrangement of Topic 15-6 if there is a limited need for shifting beam direction.

A three or four-element beam (Fig. 15-18) is manageable and not too expensive for 6-meter operation. The dimensions for the three-element type are given in Fig. 15-19. I don't know the manufacturer of this beam and I'm the third user. Shop around hamfests for bargain beams. Many are in good shape or repairable with little trouble. The beam of Fig. 15-18 has a gamma match device instead of the beta match mentioned in Topic 15-4.

Fig. 15-18. A 6-meter 3-element beam.

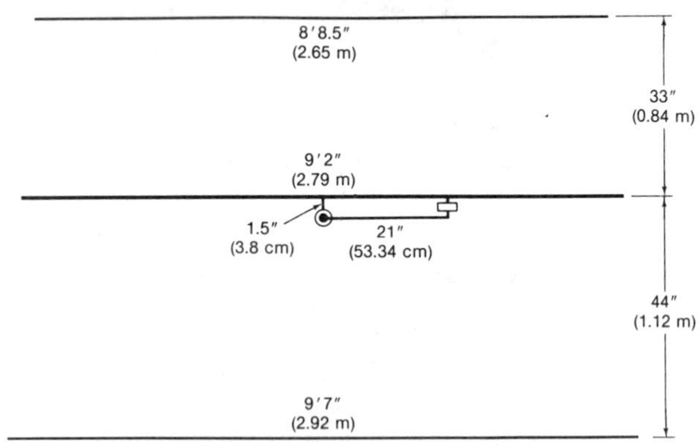

Fig. 15-19. Dimensions of 6-meter beam.

(A) 70-cm beam.

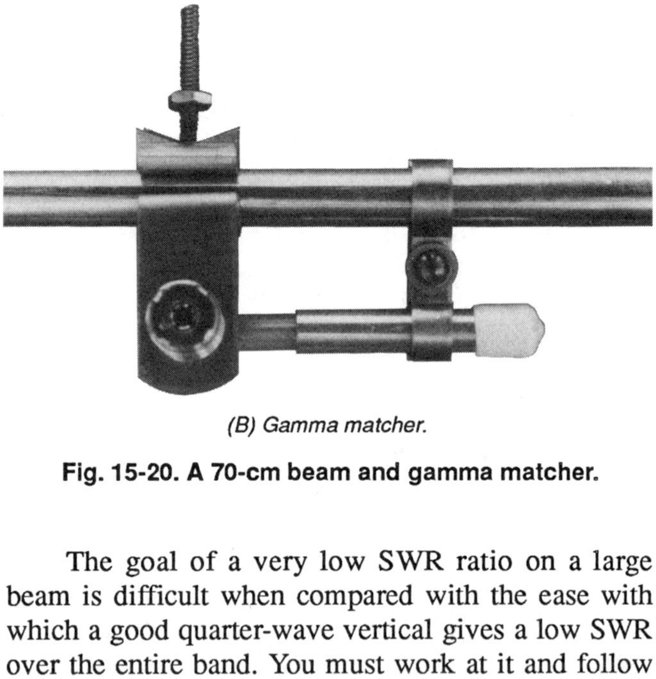

(B) Gamma matcher.

Fig. 15-20. A 70-cm beam and gamma matcher.

The Cushcraft A449-6 six-element beam of Fig. 15-20 is an inexpensive one which produces a gain of 10 dB despite its boom length of only 3 ft (.91 m). Additional dimensions are given in Fig. 15-21. Compare them with those of the three element 6-meter beam to recognize just how different dimensions are for the various VHF/UHF bands.

Most purchased beams must be assembled. Do the job exactly. Many beams have a narrow bandwidth and some of the element lengths must be adjusted rather finely to peak performance on a chosen sector of a band. Sometimes a manufacturer will have more than one model for a given band or you can have the antenna pretuned according to your preference. Take this into consideration when you purchase a very high-gain many-element beam. Bandwidth becomes much more of a limiting factor as the number of elements is increased.

The goal of a very low SWR ratio on a large beam is difficult when compared with the ease with which a good quarter-wave vertical gives a low SWR over the entire band. You must work at it and follow instructions carefully as presented in the installation procedures. Nearby antennas, power lines, and metallic surfaces can be troublesome, especially when you have a less than perfect match at the antenna. Often conditions can be improved if you move the antenna in the clear as much as possible because some matching systems are influenced by the metallic surroundings.

Most beams can be checked out on reception once you have established a reasonable SWR ratio. In most locations there are repeaters at varying distances

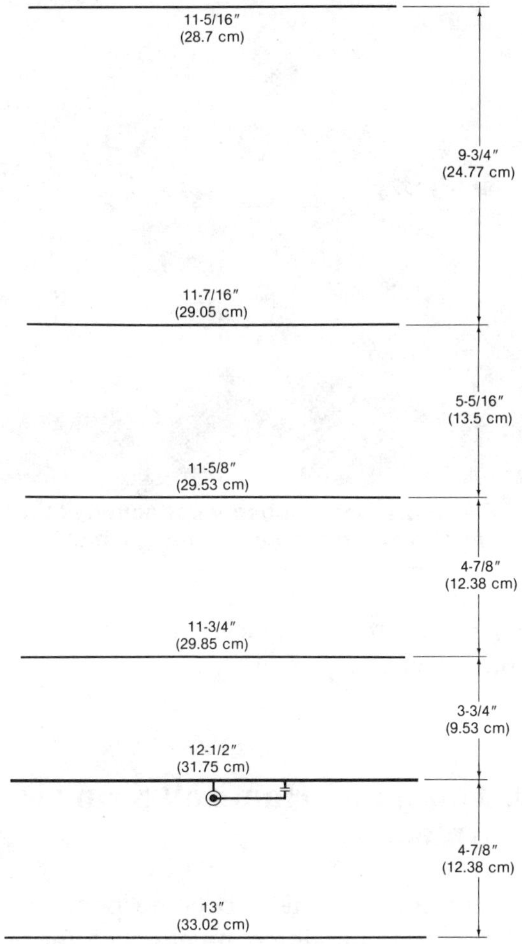

Fig. 15-21. Typical dimensions for a commercial 70-cm beam.

and you can check the received signals by comparing them with that received by an omnidirectional reference antenna. Also you can obtain a good idea of the back and side pickup when compared with the reference antenna. After that you can go ahead and get some on-the-air comparisons. You must use a good quality UHF coaxial switch when making antenna comparisons. Otherwise, you must manually change the cable. A repeater signal is useful when this is done because it is likely to change very little as a result of propagation variations while making the changeover. Of course, testing with a mobile signal when it is in motion can be more difficult because of the many variables that can influence the propagation path at such low height. You can do much better checking antennas with fixed installations. Mount your antennas in a reasonably clear location and choose a reasonable distance separation.

Fig. 15-22. Austin duobander on car roof.

15-8. Austin 6-Meter/70-Centimeter Duobander

Austin Custom Antenna has applied their ingenious decoupling systems in producing a series of mobile and base station antennas for radio hams and scanner enthusiasts. An economical and good performer is the ham duoband 2-meter/70-centimeter mobile unit of Fig. 15-22. The antenna is used in both applications and consists of an end-fed half wave 70-cm section at the top while the combined top and base sections form a quarter-wave vertical on 2 meters. The end feed of the vertical and the decoupling of the sections is handled by a coaxial cavity.

The antenna can be screwed into a versatile magnet mount that can also serve as a mount for scanner versions of the same antenna as well as landmobile and marine two-way radio antennas. The photograph of Fig. 15-23 shows several Austin antennas and the magnet mount and demonstrates the versatility of the

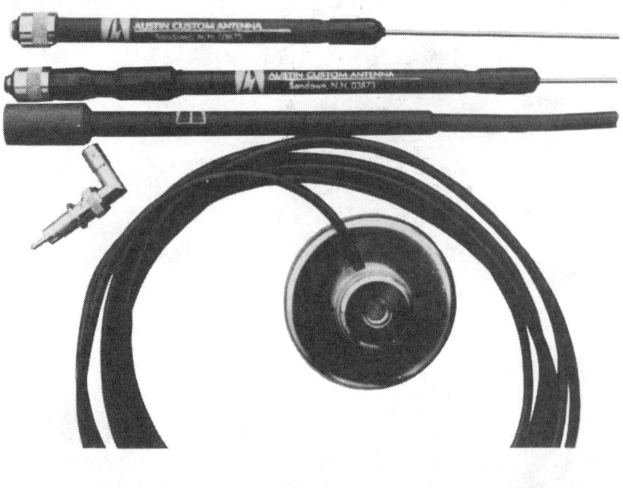

Fig. 15-23. Components of versatile Austin magnet mount and exchangeable multiband antennas.

Fig. 15-24. The Duobander attached to the Austin ground-plane right-angle mount.

idea. The second antenna from the top in Fig. 15-23 is the ham 2-meter/70-cm duobander. At the top is a similar antenna useful for the VHF and UHF landmobile radio services. The bottom antenna is a flexible one also designed for the landmobile services. This latter unit can be operated on a portable scanner using its BNC male connector or with the magnet mount by way of an included BNC-to-magnet mount adaptor. An available elbow adaptor also permits you to connect the same antenna directly to the back of a fixed scanner.

The ham duoband antennas as well as certain scanner models also fit the Austin GP-1 ground mount of Fig. 15-24. A right angle mount is supplied for side mounting or bolting to a metal or PVC mast. Austin has a three- band vertical available for the 2-meter/1.25-meter/70-centimeter and is fed with a single coaxial cable.

15-9. Multiband Ham and Scanner Antennas

Multiband antennas have become popular in the VHF/UHF spectrum. Some are designed for the ham bands; others only for reception on the landmobile, marine, VHF aviation, and public radio services. These are designed for the scanner enthusiasts. The new innovation for mast and car top antennas is the extremely wideband unit that covers all the VHF/UHF radio services and can be used for both transmit and

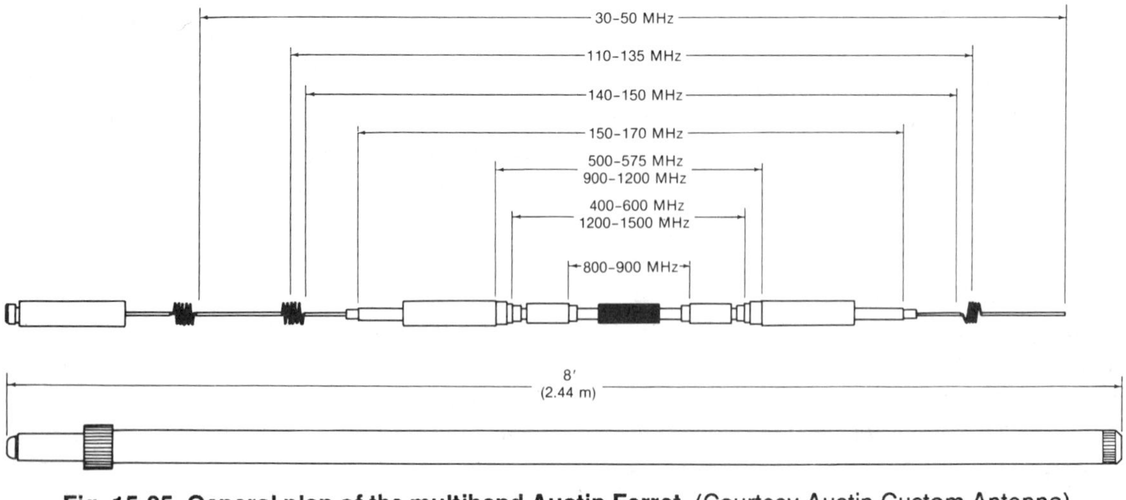

Fig. 15-25. General plan of the multiband Austin Ferret. (Courtesy Austin Custom Antenna)

receive. These antennas include discones and their variations, as well as staff antennas composed of multituned resonant segments, dielectrically tuned chokes, and other cavity couplers.

The Austin Ferret of Fig. 12-25 is an example of the cavity-coupler design. This antenna has a receive frequency coverage from 30 to 1300 MHz. For reception purposes only, a 10-dB gain antenna-mounted preamplifier is available with the power supplied via the coaxial line. For transmit, operations are centered about 40, 100, 122, 146, 155, 223, 310, 465, 865, and 1200 MHz. This four pound 8-ft antenna is mounted within a lightweight fiberglass cover. An aluminum mounting tube provides convenient support to the mast or other side brackets. The antenna terminates in an N-connector. Note in Fig. 15-25 how its operation is segmented about its center with appropriate decoupling among various frequency spectra. No radials are required. Upon request the Ferret can be purchased with responses centered in the VHF/UHF ham bands.

The discone receive operation was described in Chapter 10. As mentioned previously it has ham transmit capability on 50, 144, 430, 904, and 1200 MHz. On 50 MHz, however, a tuner was used to bring down the SWR to a more acceptable level. also it was possible to load the discone on 10 meters as well as the CB band with the use of a tuner to obtain a low SWR.

In summary, antennas like the discone and Ferret combine scanner and ham radio operations in a single package. A substantial number of radio amateurs are interested in and enjoy both activities. Likewise scanner fans like to listen to activities on the ham bands.

Chapter 16

International Antenna Dimension Tables for Ham Radio Operators

THE choices of frequencies presented in the amateur radio dimension tables have been selected to accommodate worldwide assignments. Bands, even segments of individual bands, are not the same in all countries. It is your responsibility to know the exact frequencies that are permitted in your country. If we have overlooked any active frequency assignment in some part of the world contact us for inclusion of the proper data in the next revision.

Table 16-1 covers the amateur assignments between 1.8 and 30 MHz. Dimensions are given in feet and meters. Table 16-2 presents the VHF/UHF assignments. Dimensions are in inches and centimeters. Recall also the useful information given in Tables 11-4, 11-5, and 11-6 when making conversions or in the cuttings of antennas to some precise frequency not given in the table.

You will note that one frequency in many of the listed bands has been marked with a dagger (†). These frequencies are popular for single-sideband operation. Many of the antennas built for test application during book preparation were dimensioned to these frequencies.

The minimum SWR does not always occur on the precise cut frequency. Sometimes there was some measurement slip or oversight on our part. However you should not always expect a perfect alignment of cut and resonant frequency because of the variables of site, conditions, height, materials, insulators, connectors, etc. Expect to do some trimming and adjusting in almost every case if you are bent on obtaining the very minimum SWR on some precise frequency. Fortunately such an undertaking is often quite unnecessary because most of the antennas presented have a reasonable bandwidth.

Table 16-1. Amateur HF Band Dimensions*

Freq. (MHz)	Meters	λ/4 Element	λ/4 Radial	3/4λ Element	Reflector & Free Space λ	Director	Beam Driven Element	Free Space λ	Free Space 0.25 λ	Free Space 0.2 λ	λ/4 Stub 0.66 VF	λ/4 Stub 0.79 VF
160 Meters												
1810	166	129'	133'	401'	272'	246'	133'	544'	136'	109'	90'	107'
		39.4	40.4	122	82.9	75.1	40.6	166	41.4	33.1	27.3	32.7
1830	164	128'	131'	397'	269'	243'	132'	538'	134'	108'	88'8"	106'
		39.0	40.0	121	82.0	74.3	40.2	164	41	32.8	27.0	32.4
1850†	162	126'	130'	392'	266'	241'	130'	532'	133'	106'	87'9"	105'
		38.5	39.5	120	81.1	73.5	39.7	162	40.5	32.4	26.8	32.0

Table 16-1 — cont. Amateur HF Band Dimensions*

Freq. (MHz)	Meters	λ/4 Element	λ/4 Radial	3/4λ Element	Reflector & Free Space λ	Director	Beam Driven Element	Free Space λ	Free Space 0.25 λ	Free Space 0.2 λ	λ/4 Stub 0.66 VF	λ/4 Stub 0.79 VF
1870	160	125' 38.1	128' 39.1	388' 118	263' 80.2	238' 72.7	129' 39.3	526' 160	132' 40.1	105' 32.1	86'10" 26.5	104' 31.7
1890	159	124' 37.7	127' 38.7	384' 117	260' 79.4	235' 72.0	128' 38.9	521' 159	130' 39.7	104' 31.7	85'11" 26.2	103' 31.3
1910	157	123' 37.3	126' 38.3	380' 116	258' 78.5	233' 71.2	126' 38.5	515' 157	129' 39.3	103' 31.4	85' 25.9	102' 31.0
1930	155	121' 36.9	124' 37.9	376' 115	255' 77.7	231' 70.5	125' 38.1	510' 155	127' 38.9	102' 31.1	84'2" 25.6	101' 30.7
1950	154	120' 36.6	123' 37.5	372' 113	252' 76.9	228' 69.7	124' 37.7	505' 154	126' 38.5	101' 30.8	83'3" 25.4	99'8" 30.4
1970	152	119' 36.2	122' 37.1	369' 112	250' 76.1	226' 69.0	122' 37.3	499' 152	125' 38.1	100' 30.5	82'5" 25.1	98'8" 30.1
1990	151	118' 35.8	121' 36.8	365' 111	247' 75.4	224' 68.3	121' 36.9	494' 151	124' 37.7	98'11" 30.1	81'7" 24.9	97'8" 29.8
80 Meters												
3520	85.2	66'6" 20.3	68'2" 20.8	206' 62.9	140' 42.6	126' 38.6	68'6" 20.9	280' 85.2	69'11" 21.3	55'11" 17.0	46'2" 14.1	55'2" 16.8
3560	84.3	65'9" 20.0	67'5" 20.5	204' 62.2	138' 42.1	125' 38.2	67'8" 20.0	276' 84.3	69'1" 21.1	55'3" 16.9	45'7" 13.9	54'7" 16.6
3600	83.3	65' 19.8	66'8" 20.3	202' 61.5	137' 41.7	124' 37.8	66'11" 20.4	273' 83.3	68'4" 20.8	54'8" 16.7	45'1" 13.7	54' 16.5
3640	82.4	64'4" 19.6	65'11" 20.1	199' 60.8	135' 41.2	122' 37.4	66'2" 20.2	270' 82.4	67'7" 20.6	54'1" 16.5	44'7" 13.6	53'5" 16.3
3680	81.5	63'7" 19.4	65'3" 19.9	197' 60.1	134' 40.8	121' 37.0	65'6" 20.0	267' 81.5	66'10" 20.4	53'6" 16.3	44'2" 13.4	52'10" 16.1
3720	80.6	62'11" 19.2	64'6" 19.7	195' 59.5	132' 40.3	120' 36.6	64'9" 19.8	265' 80.6	66'2" 20.2	52'11" 16.1	43'8" 13.3	52'3" 15.9
3800†	79	61'7" 18.8	63'2" 19.3	191' 58.2	129' 39.5	117' 35.7	63'5" 19.3	259' 78.9	64'9" 19.7	51'10" 15.8	42'9" 13.0	51'2" 15.6
3840	78	60'11" 18.6	62'6" 19.0	189' 57.6	128' 39.1	116' 35.4	62'9" 19.1	256' 78.1	64' 19.5	51'3" 15.6	42'3" 12.9	50'7" 15.4
3880	77	60'4" 18.4	61'10" 18.9	187' 57.0	127' 38.7	115' 35.0	62'1" 18.9	254' 77.3	63'5" 19.3	50'9" 15.5	41'10" 12.8	50'1" 15.3
3920	76.5	59'8" 18.2	61'3" 18.7	185' 56.5	126' 38.3	114' 34.7	61'6" 18.8	251' 76.5	62'9" 19.1	50'2" 15.3	41'5 " 12.6	49'7" 15.1
3960	76	59'1" 18.0	60'7" 18.5	183' 55.9	124' 37.9	112' 34.3	60'10" 18.6	248' 75.8	62'1" 18.9	49'8" 15.2	41' 12.5	49'1" 15.0
40 Meters												
7020	42.7	33'4" 10.2	34'2" 10.42	103' 31.5	70' 21.4	63'5" 19.4	34'3" 10.47	140' 42.7	35' 10.7	28' 8.55	23'2" 7.05	27'8" 8.44
7060	42.5	33'2" 10.1	34'0" 10.36	102'10" 31.3	69'8" 21.2	63' 19.3	34'2" 10.4	139' 42.5	34'10" 10.6	27'11" 8.50	23' 7.01	27'6" 8.39
7100	42.3	33' 10.04	33'10" 10.3	102'3" 31.2	69'4" 21.1	62'8" 19.2	33'11" 10.35	138'7" 42.3	34'8" 10.56	27'9" 8.45	22'10" 6.97	27'4" 8.34
7140	42	32'9" 10.0	33'7" 10.25	101'8" 31.0	68'11" 21.0	62'4" 19.0	33'9" 10.3	137'10" 42.0	34'5" 10.5	27'7" 8.4	22'9" 6.93	27'2" 8.30
7180	41.8	32'7" 9.93	33'5" 10.2	101'1" 30.8	68'6" 20.9	62' 18.9	33'7" 10.2	137'1" 41.8	34'3" 10.45	27'5" 8.36	22'7" 6.89	27'1" 8.25
7200†	41.7	32'6" 9.9	33'4" 10.16	100'10" 30.7	68'4" 20.8	61'10" 18.9	33'6" 10.21	136'8" 41.7	34'2" 10.4	27'4" 8.33	22'7" 6.88	27' 8.23
7240	41.4	32'4" 9.85	33'2" 10.1	100'3" 30.6	68' 20.7	61'6" 18.8	33'4" 15.52	135'11" 41.4	34' 10.36	27'2" 8.29	22'5" 6.84	26'10" 8.18
7280	41.2	32'2" 9.80	33' 10.0	99'9" 30.4	67'7" 20.6	61'2" 18.7	33'1" 10.1	135'2" 41.2	33'10" 10.3	27' 8.24	22'4" 6.80	26'8" 8.14

Table 16-1 — cont. Amateur HF Band Dimensions*

Freq. (MHz)	Meters	λ/4 Element	λ/4 Radial	3/4λ Element	Reflector & Free Space λ	Director	Beam Driven Element	Free Space λ	Free Space 0.25 λ	Free Space 0.2 λ	λ/4 Stub 0.66 VF	λ/4 Stub 0.79 VF
30 Meters												
10120	29.6	23'1" 7.05	23'8" 7.23	71'8" 21.9	48'7" 14.8	44' 13.4	23'10" 7.26	97'3" 29.64	24'4" 7.44	19'5" 5.93	16' 4.89	19'2" 5.85
20 Meters												
14060	21.3	16'8" 5.071	17'1" 5.203	51'8" 15.74	35' 10.67	31'8" 9.673	17'2" 5.228	70' 21.34	17'6" 5.334	14' 4.267	11'7" 3.52	13'10" 4.213
14180	21.16	16'6" 5.03	16'11" 5.159	51'2" 15.61	34'8" 10.58	31'5" 9.591	17' 5.183	69'5" 21.16	17'4" 5.289	13'11" 4.231	11'5" 3.49	13'8" 4.177
14240†	21.1	16'5" 5.01	16'10" 5.137	51' 15.54	34'7" 10.53	31'3" 9.551	16'11" 5.162	69'1" 21.07	17'3" 5.267	13'10" 4.213	11'5" 3.475	13'8" 4.159
14300	21.0	16'4" 4.986	16'9" 5.115	50'9" 15.48	34'5" 10.49	31'1" 9.51	16'10" 5.14	68'10" 20.98	17'2" 5.245	13'9" 4.196	11'4" 3.461	13'7" 4.142
17 Meters												
18110	16.6	12'11" 3.937	13'3" 4.039	40'1" 12.22	27'2" 8.283	24'7" 7.510	13'4" 4.059	54'4" 16.57	13'7" 4.141	10'10" 3.313	9' 2.733	10'9" 3.271
15 Meters												
21040	14.3	11'1" 3.389	11'5" 3.477	34'6" 10.52	23'5" 7.129	21'2" 6.464	11'5" 3.493	46'9" 14.26	11'8" 3.565	9'4" 2.852	7'9" 2.352	9'3" 2.815
21120	14.2	11'1" 3.376	11'4" 3.464	34'5" 10.48	23'4" 7.102	21'1" 6.439	11'5" 3.480	46'7" 14.21	11'8" 3.551	9'4" 2.841	7'8" 2.343	9'2" 2.804
21300†	14.1	11' 13.347	11'3" 3.434	34'1" 10.39	23'1" 7.042	20'11" 6.385	11'4" 3.451	46'2" 14.09	11'7" 3.521	9'3" 2.817	7'7" 2.323	9'1" 2.781
21400	14	10'11" 3.332	11'2" 3.418	33'11" 10.34	23' 7.009	20'10" 6.355	11'3" 3.435	46' 14.02	11'6" 3.505	9'2" 2.804	7'7" 2.313	9'1" 2.768
12 Meters												
24.930	12	9'5" 2.860	9'8" 2.934	29'1" 8.877	19'9" 6.015	17'10" 5.455	9'8" 2.948	39'6" 12.03	9'10" 3.008	7'11" 2.407	6'6" 1.985	7'10" 2.376
10 Meters												
28100	10.78	8'4" 2.537	8'7" 2.603	25'10" 7.875	17'6" 5.338	15'10" 4.840	8'7" 2.616	35' 10.68	8'9" 2.669	7' 2.135	5'9" 1.761	6'11" 2.108
28500†	10.5	8'2" 2.502	8'5" 2.567	25'6" 7.765	17'3" 5.263	15'7" 4.772	8'6" 2.579	34'6" 10.53	8'8" 2.632	6'11" 2.105	5'8" 1.737	6'10" 2.078
29000	10.35	8'1" 2.459	8'3" 2.522	25' 7.631	17' 5.172	15'4" 4.690	8'4" 2.534	33'11" 10.35	8'6" 2.586	6'10" 2.069	5'7" 1.707	6'8" 2.042
29500	10.17	7'11" 2.417	8'2" 2.480	24'7" 7.502	16'8 " 5.085	15'1" 4.61	8'2" 2.492	33'4" 10.17	8'4" 2.542	6'9" 2.034	5'6" 1.678	6'7" 2.008

*Dimensions are in feet on first line and meters on second line.
†Frequencies popular for single-sideband operation.

Table 16-2. VHF/UHF Amateur Band Dimensions*

Freq. (MHz)	Meters	λ/4 Element	λ/4 Radial	3/4λ Element	Reflector & Free Space λ	Director	Beam Driven Element	Free Space λ	Free Space 0.25 λ	Free Space 0.2 λ	λ/4 Stub 0.66 VF	λ/4 Stub 0.79 VF
6 Meters												
50.3†	5.96	55.1	57.3	172	116	106	56.7	235	58.69	46.96	38.73	46.36
		140	145	438	294	269	144	596.4	149.1	119.3	98.38	117.76
51.5	5.83	53.8	55.9	168	113	103	55.4	229	57.32	45.86	37.83	45.28
		137	142	428	287	262	141	582.5	145.6	116.5	96.09	115.0
52.5FM	5.71	52.8	54.9	165	111	101	54.3	225	56.23	44.99	37.11	44.42
		134	139	420	282	257	138	571.4	142.9	114.3	94.27	112.9
4 Meters												
70.3	4.27	39.40	40.97	123	82.8	75.7	40.6	168	42.0	33.6	27.7	33.2
		100	104	313	210	192	103	427	107	85.3	70.4	84.3
2 Meters												
144.2†	2.08	19.2	20.0	60.2	40.4	36.9	19.79	81.89	20.47	16.38	13.51	16.17
		48.8	50.7	153	103	93.7	50.26	208	52.01	41.61	34.32	41.08
145	2.07	19.1	19.9	59.8	40.14	36.7	19.68	81.43	20.36	16.29	13.43	16.08
		48.5	50.4	152	102	93.2	49.98	207	51.72	41.38	34.13	40.85
146FM	2.05	19.0	19.7	59.4	39.9	36.4	19.54	80.88	20.22	16.18	13.34	15.97
		48.2	50.1	151	101	92.6	49.64	205	51.37	41.10	33.90	40.57
147	2.04	18.8	19.6	59.0	39.6	36.2	19.41	80.33	20.08	16.07	13.25	15.86
		47.9	49.8	150	100.6	91.9	49.30	204	51.02	40.82	33.67	40.30
1.25 Meters												
221	1.36	12.53	13.03	39.26	26.34	24.07	12.91	53.43	13.36	10.69	8.814	10.55
		31.84	33.10	99.72	66.89	61.15	32.79	135.7	33.94	27.15	22.34	26.80
222†	1.35	12.48	12.97	39.08	26.22	23.96	12.85	53.19	13.30	10.64	8.775	10.51
		31.69	32.95	99.27	66.59	60.87	32.64	135.1	33.78	27.03	22.24	26.69
223	1.34	12.42	12.92	38.91	26.10	23.86	12.79	52.95	13.24	10.59	8.735	10.46
		31.55	32.80	98.82	66.29	60.60	32.50	134.5	33.63	26.91	22.19	26.56
70 Centimeters												
425	0.705	6.518	6.776	20.41	13.69	12.52	6.713	27.78	6.946	5.558	4.584	5.487
		16.56	17.21	51.85	34.78	31.79	17.05	70.59	17.65	14.12	11.64	13.94
430	0.697	6.442	6.698	20.18	13.54	12.37	6.635	27.46	6.865	5.493	4.530	5.423
		16.36	17.01	51.25	34.38	31.43	16.85	69.77	17.44	13.95	11.51	13.78
435	0.689	6.368	6.621	19.95	13.38	12.23	6.559	27.15	6.786	5.430	4.478	5.361
		16.18	16.82	50.66	33.98	31.06	16.66	68.97	17.24	13.79	11.38	13.62
440	0.681	6.295	6.545	19.72	13.23	12.09	6.484	26.84	6.709	5.368	4.427	5.300
		15.99	16.63	50.08	33.60	30.77	16.47	68.18	17.05	13.64	11.25	13.46
445	0.674	6.225	6.472	19.50	13.08	11.96	6.411	26.54	6.634	5.308	4.378	5.240
		15.81	16.44	49.52	33.22	30.37	16.29	67.42	16.85	13.48	11.12	13.31

*Dimensions are in inches on first line and centimeters on second line.
†Frequencies popular for single-sideband operation.

Appendix

Antenna Manufacturers, Sources, and Information

Amateur Electronic Supply
4828 W. Fond du Lac Ave.
Milwaukee, WI 53216

Antenna Specialists
12435 Euclid Ave.
Cleveland, OH 44106

Austin Custom Antenna
P.O. Box 357
Sandown, NH 03873

Celwave R.F. Inc.
Route 79
Marlboro, NJ 07746

Cushcraft Communication Antennas
P.O. Box 4680
Manchester, NH 03108

Encomm
1506 Capital Ave.
Plano, TX 75074

Grove Enterprises
P.O. Box 98
Brasstown, NC 28902

MFJ Enterprises
P.O. Box 494
Mississippi State, MS 39762

Noll Ed
P.O. Box 1042
Doylestown, PA 18901

Radio Publications, Inc.
P.O. Box 149
Wilton, CT 06897

RF Electronics
1164 S. Harbor Blvd.
Fullerton, CA 92632

Radio West
850 Anns Way
Vista, CA 92083

Telex Communications (HyGain)
9600 Aldrich Ave. S.
Minneapolis, MN 55420

W9INN Antennas
P.O. Box 393
Mt. Prospect, IL 60056

Index

I

L

M

N

O

P

Q

R

S

T

U

V

W

Y

Z

Landmobile and Marine Radio Technical Handbook
Edward M. Noll

A must for anyone interested in radio communications, this book is an invaluable learning tool for two-way radio communications and a complete reference source for experienced users.

It covers the entire two-way field-private landmobile service, marine radio telephone and radiotelegraph, marine navigation, and Citizens Band. And if you're seeking certification, it's an ideal study guide, reflecting up-to-date FCC licensing information.

From the fundamentals through advanced topics, *Landmobile and Marine Radio Technical Handbook* will explain everything you need to know about two-way radio—and then some.

Topics covered include:

- Two-Way Radio Services
- Transmission Characteristics
- Solid-State Fundamentals
- Modulation Systems
- Digital and Microprocessor Electronics
- Test Equipment and Methods
- Antenna Systems and Interference
- Landmobile Radio Equipment
- Repeater, Trunked, and Cellular Radio Systems
- Marine Radio and Navigational Systems for Small Craft
- Marine Radio Systems for Larger Craft
- Marine Direction Finders
- Marine Radar
- FCC Licensing, Operating Practices, and Examination Requirements

576 Pages, 7½ x 9¾, Softbound
ISBN: 0-672-22427-5
No. 22427, $24.95

Shortwave Radio Listening with the Experts
Gerry L. Dexter, Editor

Experts in shortwave listening share their expertise to make this single-volume treasure complete. Its 25 chapters cover everything needed to get the most value from both the simplest and the most sophisticated receiver.

Because even the more experienced listeners are not equally strong in all facets of band specialties, this book can be used by the beginner as well as the established listener.

Topics covered include:

- Setting Up Your Shack
- Antennas for Shortwave Reception
- Purchasing the Right Shortwave Receiver
- Fundamentals of Shortwave Radio Propagation
- Recognizing Languages
- Printed Materials for the DX'er
- Tape Recording and Shortwave
- The Verification Game
- The Association of North American Radio Clubs
- Computers and DX'ing
- Over-the-Horizon Radar
- DX'ing the U.S. Pirates
- An Overview of the Numbers Stations Mystery
- Challenges and Changes in Medium-Wave Listening
- FM DX'ing
- Exploring the VHF/UHF Action Bands
- Utility Monitoring
- World Radioteletype Press Stations
- Amateur Band DX'ing

The 25 contributors are known by the shortwave listening community for their area of expertise.

528 Pages, 7½ x 9¾, Softbound
ISBN: 0-672-22519-0
No. 22519, $22.95

World Satellite Almanac, Second Edition
Mark Long

This book is a complete guide to the transmission and reception of video, voice, and data communications technology. The second edition is expanded to provide a global perspective for the developing technologies in satellite communications.

Workers in the satellite field will welcome this comprehensive reference that contains a wealth of charts, maps, tables, and graphs relating to satellite footprints, as well as current information regarding FCC decisions.

Topics covered include:

- Evolution of Satellite Communications
- World Video Standards
- Satellite Voice and Data Transmission Techniques
- International Satellite Reception Techniques
- The Intelsat Global Satellite Communication System
- Intersputnik and the Soviet Statsionar Satellite System
- The Inmarsat Maritime Satellite System
- Eutelsat and the European Satellite System
- Arabsat Satellite System
- Palapa Satellite System
- International Telecommunications Union (ITU) Regions 1,2, and 3 (within each region, lists every satellite by location; includes technical specifications, channel assignments, and footprint maps.)

Mark Long has written several best-selling books and over 100 magazine and newspaper articles on satellite technology.

704 Pages, 7½ x 9¾, Softbound
ISBN: 0-672-22559-X
No. 22559, $34.95

Reference Data for Engineers: Radio, Electronics, Computer, and Communications, Seventh Edition
Edward C. Jordan, Editor-in-Chief

Written by 70 professionals, this book was originally published as a limited private edition used by engineers, mathematicians, and physicists at ITT. Forty years later, it has become a familiar and dependable reference for engineers worldwide.

In this completely updated seventh edition, the title has been changed to reflect the range of disciplines included. The scope of coverage has been greatly expanded.

Topics covered include:

- International Telecommunication Recommendations
- Units, Constants, and Conversion Factors
- Filters, Simple Band-Pass Design
- Modern-Network-Theory Design
- Rectifiers, Filters, and Power Supplies
- Optical, Analog, and Digital Communications
- Computer Communications Networks
- Satellite and Space Communications
- Broadcasting, Cable Television, and Recording System Standards
- Switching Networks and Traffic Concepts
- Electroacoustics
- Computer Organization and Programming

This expanded edition contains 48 chapters edited by Edward C. Jordan and written by more than 70 engineers, scientists, educators, and specialists.

1,368 Pages, 7½ x 9¾, Hardbound
ISBN: 0-672-21563-2
No. 21563, $79.95

Visit your local book retailer or call 800-428-SAMS.

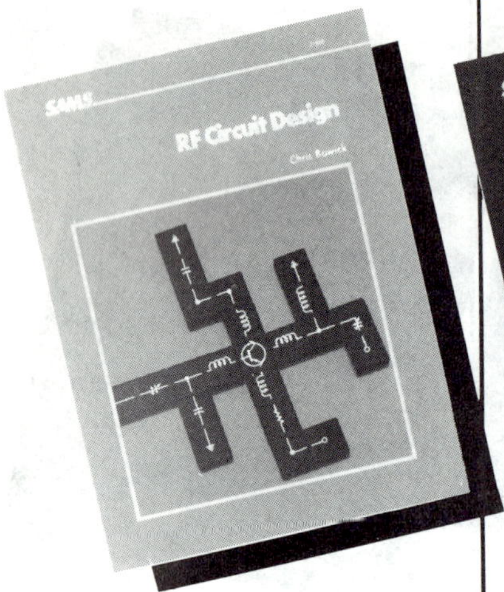

RF Circuit Design
Christopher J. Bowick

Enjoy the benefits of two books in one. Use this in cookbook fashion as a catalog of useful circuits or as a reference manual. It clearly presents a user-oriented approach to the design of RF amplifiers and impedance matching networks and filters.

The book teaches procedures for designing your own low-pass, high-pass, bandpass, and bandstop filter designs; how to design single and multiresonator circuits; and the difference between small- and large-signal amplifiers and the step-by-step procedures for designing them.

Topics covered include:

- Components
- Resonant Circuits
- Filter Design
- Impedance Matching
- The Transistor at Radio Frequencies
- Small-Signal RF Amplifier Design
- R Power Amplifiers
- Appendices: Vector Algebra, Noise Calculations, Bibliography

176 Pages, 8½ x 11, Softbound
ISBN: 0-672-21868-2
21868, $22.95

Commodore 64 and 128 Programs for Amateur Radio & Electronics
Joseph J. Carr

The electronics hobbyist, programmer, engineer, and technician will enjoy the task-oriented programs for amateur radio and electronics in this book.

The book contains two general categories of programs—amateur radio technology and general electronics—that will simplify programming tasks when incorporated into the custom-designed software programs provided. It also makes full use of the analytical potential of the personal computer with 42 practical programs including amateur radio programs.

Topics covered include:

- Antenna Calculations
- Standing-Wave Ratio
- Transmission Line
- Impedance Matching

- Satellite Dish Antenna
- Resistors in Series or Parallel
- Capacitors in Series or Parallel
- Inductors in Series or Parallel
- RLC Networks
- Operational Amplifier Circuits

Joseph Carr, an electrical engineer, has written over 43 books and more than 200 magazine articles.

256 Pages, 5½ x 8½, Softbound
ISBN: 0-672-22516-6
No. 22516, $14.95

Mobile Communications Design Fundamentals
William C.Y. Lee

An authoritative introduction to mobile communications design, this thought-provoking professional reference provides easy-to-understand methods of analysis for the communications engineer and the engineering student.

Mobile Communications Design Fundamentals provides many tools a system designer will need.

Practical design methods and curves are presented in a straightforward format, using many helpful illustrations and examples. This text avoids heavy mathematics, relying instead on physical interpretations that are easier to read and faster to learn.

Topics covered include:

- The Mobile Radio Environment
- Prediction of Propagation Loss
- Calculation of Fades and Methods of Reducing Fades
- Mobile Radio Interference
- Frequency Plans and Their Associated Schemes
- Design Parameters at the Mobile Unit
- Signaling and Channel Access
- Miscellaneous Related Systems

Dr. William C.Y. Lee, a pioneer in mobile radio communications, has worked with advanced programs for various military systems and has conducted research in satellite communications.

352 Pages, 7½ x 9¾, Hardbound
ISBN: 0-672-22305-8
No. 22305, $34.95

Telecommunications Networks: A Technical Introduction
Robert J. "Bert" Murphy

Rapid technical advances and the increased importance of transmitting information (voice and data) have created a need for documented technical information that, until now, was found only in industry journals.

This comprehensive introduction to telecommunications networks offers an overview for the telecommunications or data communications professional. It provides especially strong chapters on switching, traffic engineering, transmission, ISDN, and network management.

Topics covered include:

- History of the Information Age
- Switching Systems
- Traffic Engineering
- Transmission
- Data
- Data Systems Configuration
- Telecommunication Services
- Current Network Environment
- Integrated Services Digital Network
- Network Management
- Planning a Private Network
- Network Illustrations
- Future Networks

336 Pages, 7½ x 9¾, Hardbound
ISBN: 0-672-22588-3
No. 22588, $29.95

Visit your local book retailer or call 800-428-SAMS.

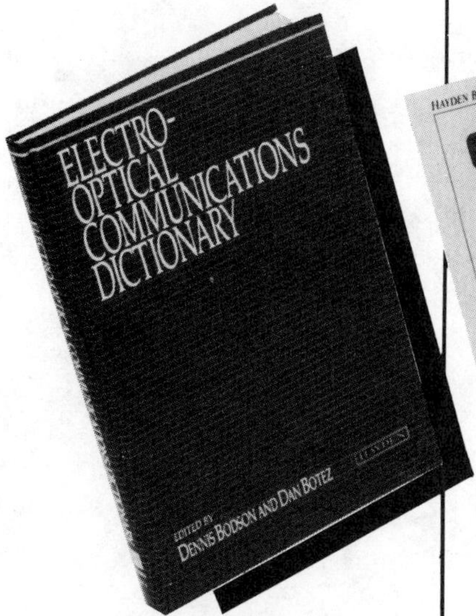

Electro-Optical Communications Dictionary

Bodson and Botez, Editors

Advances in laser, computer, process, and manufacturing technologies have all contributed greatly to the growth and development of the electro-optical communications field. This reference enables users with those varying backgrounds to achieve understanding and use the precise terminology associated with electro-optical communications.

Terms and definitions presented in this dictionary are consistent with federal, industry, and professional society standards. Processes, devices, and components usually found and used in electro-optical communications are clearly described. Cross-references and inversions are utilized throughout to provide quick and easy access, and a list of acronyms and abbreviations often encountered in this emerging field is provided.

This comprehensive reference guide contains over 5,000 terms and definitions useful to anyone involved with electro-optical communication systems.

Topics covered include:

- Fiber-Optic Components
- Electromagnetic Transmission Systems
- Optical/Electrical Interfacing
- Optical/Electromagnetic Theory
- Optical/Digital/Analog Systems
- Information Theory
- Lightwave Communications

176 Pages, 6¼ x 9¼, Hardbound
ISBN: 0-8104-0961-5
No. 40961, $24.95

Radio Operator's License Question & Answer Manual Eleventh Edition

Milton Kaufman

This classic best-seller is a complete guide to radio communications and a valuable resource for anyone wanting to pass the FCC licensing exam. The newly revised 11th edition updates FCC rules and regulations, includes sample test questions, and provides a "practice" licensing exam with typical multiple-choice questions.

Topics covered include:

- Basic Law
- Basic Operating Practice
- Basic Radiotelephone
- Alternating and Direct Current, Electron Tubes, and Power Supplies
- Indicating Instruments and Oscillators
- Audio, RF, IF, AGC, and Operational Amplifiers
- Transmitters and Mobile Radio and TV
- Amplitude and Frequency Modulation
- Diodes, Transistors, and Digital Principles, Circuits, and Test Equipment
- Antennas and Transmission Lines
- Frequency Measurements and Radio Frequency Interference
- Land Mobile Interference
- Interference to Home Entertainment Equipment
- Batteries, Motors and Generators, Microwave Equipment
- Rules and Regulations
- Ship Radar Techniques

550 Pages, 7½ x 9¾, Softbound
ISBN: 0-672-48444-7
No. 48444, $19.95

Basic Carrier Telephony, Third Edition

David Talley

This text is a basic course in the principles and applications of carrier telephony. Abundantly illustrated, with questions and problems throughout, and requiring little mathematics, this text has successfully trained personnel throughout the communications and electronics fields.

Learn basic carrier principles, plus modulation/demodulation processes, transmission and signaling features, and carrier applications to radio systems.

Topics covered include:

- The Telephone System
- Telephone Carrier Elements
- Wire Transmission Principles
- Telephone Transmission Practices
- FDM Carrier Systems Fundamentals
- FDM Modulation and Demodulation Process
- Electrical Filters
- FDM Carrier System Operations
- Carrier Transmission and Signaling Features
- FDM Cable Carrier Systems
- PCM Carrier Systems
- Broadband Carrier Systems and Applications

224 Pages, 6 x 9, Softbound
ISBN: 0-8104-5848-9
No. 45848, $15.95

Radio Handbook, Twenty-Third Edition

William I. Orr, Editor

The Best Ever! This 23rd edition of the radio communication handbook gives complete documentation in the hf, vhf, and uhf spectrums. Theory, equipment design, and construction techniques are covered in detail for the advanced engineer, technician, or advanced radio amateur.

Bill Orr (W6SAI), the editor, and his staff of contributing editors— all hams with extensive technical knowledge and experience—have added a wealth of brand new material on such topics as:

- Electric Wave Filters
- Installation and Maintenance of Marine Radio Equipment
- VCR and Video Disc Player Interference
- Antennas for Mobile Operations

New construction projects have been added, including:

- High-Power hf Amplifiers
- Control and Sequencing Circuits for Amateur Stations
- High-Performance Low-Pass Speech Filter
- An Easy-to-Build CW Filter
- New Advanced Communications Receiver

Also featured:

- New Tables on Computer Drive General-Purpose Filters
- New Pi and Pi-L Network Charts for Linear Amplifiers
- New Circuit Data on hf-vhf Amplifier Design
- New Data on Sloper and Inverted-V Antennas
- New Dimensional Data for hf Yagi Beam Antennas
- And Much, Much More!

672 Pages, 7¾ x 10, Hardbound
ISBN: 0-672-22424-0
No. 22424, $29.95

Visit your local book retailer or call 800-428-SAMS.